RSC Paperbacks

POLYMERS AND THE ENVIRONMENT

GERALD SCOTT

Professor Emeritus in Chemistry
Aston University
Birmingham, UK

ROYAL SOCIETY OF CHEMISTRY

ISBN 0-85404-578-3

A catalogue record for this book is available from the British Library

Published by The Royal Society of Chemistry, Thomas Graham House, Science Park, Milton Road, Cambridge CB4 0WF, UK

For further information see our web site at www.rsc.org

Typeset in Great Britain by Vision Typesetting, Manchester
Printed by Athenaeum Press Ltd, Gateshead, Tyne & Wear

Preface

The benefits conferred on society by the development of man-made polymers has in recent years been obscured by the problem of their ultimate disposal. The purpose of this book is to redress the balance. Although packaging is still the major application for plastics, engineering polymers are increasingly gaining acceptance in high technology applications, particularly in aircraft and in motor vehicles where fuel conservation is of paramount importance. The most important properties and industrial applications of polymers are reviewed in Chapter 1. In Chapter 2 the environmental benefits of polymers are compared with traditional materials with particular reference to energy demand during manufacture and use.

Rubbers and plastics are subject to environmental degradation and consequent deterioration in mechanical performance during use. Environmental durability is thus an important contributor to materials conservation. Chapter 3 outlines the role of antioxidants and stabilisers in protecting polymers against peroxidation both during processing and in subsequent service. Because of their sensitivity to the environment, polymers are also more difficult to recycle by reprocessing than traditional materials but on the other hand polymer wastes are a potential source of energy by incineration. In Chapter 4 the behaviour of polymers in alternative recycling processes (reprocessing, incineration, pyrolysis and composting) are considered in the light of current legislation.

The concept of 'biodegradable' synthetic polymers was first proposed in the 1960s as a potential solution to the problem of plastics litter. This was initially welcomed by environmentally aware public, particularly in the USA, since it was perceived to be a possible solution to the landfill problem. Packaging producers in collaboration with corn growers were quick to respond to the public mood by adding corn starch to polyethylene. Carrier bags made from this material were claimed, without experimental evidence, to be biodegradable but the

legislative authorities in the USA categorised this behaviour as 'deceptive'. The polymer-producing industries also reacted sharply against
the idea of polymer degradability since it appeared to conflict with the
basic ethos of the industry that polymers should be made as stable as
possible. These controversies obscured the real problem which was
that neither biodegradable nor non-biodegradable polymers are suitable for landfill. Biodegradable materials give rise to methane (biogas)
and cause subsequent explosion and subsidence. Carbon-chain
packaging polymers, on the other hand, are stable in landfill but their
intrinsic energy is wasted. In due course the alternative strategy of
segregating biodegradable materials for composting and non-biodegradable materials for mechanical recycling and energy production
emerged.

The problem of litter requires quite different solutions. The escala-
ting use of plastics in packaging, in agricultural mulching films and as
a replacement for sisal in hay bailing twines is now a serious pollution
problem on the land and in the sea. Not only does plastic litter result
in visual pollution but it also presents a threat to animals and birds
and results in a reduction in agricultural productivity. Litter cannot be
economically collected and recycled by any of the accepted methods
and this has led to the introduction of polyolefins with enhanced biodegradability (Chapter 5). The bioassimilation into the natural environment of used mulching films has resulted in considerable savings to
farmers and has led to conservation of fertilisers and irrigation water.
Modified natural polymers (*e.g.* starch and cellulose) are also the basis
of 'new' biodegradable polymers and a parallel search for synthetic
hydro-biodegradable polymers has resulted in polyesters which are
already used in prostheses. It is anticipated that some of these will also
find application in waste products that end up in the sewage systems
and some may even have potential application as replacements for
traditional packaging materials, for example in garden waste bags
which can be bioassimilated into compost.

This book is intended to introduce the non-specialist reader to the
environmental benefits and limitations of polymeric materials. In particular it is hoped that, by understanding the way in which polymers
interact with the environment, legislators and those in industry and
local government who have responsibility for waste management may
be encouraged to optimise the contribution of man-made polymers for
the benefit of society. There is no doubt that environmental policy will
become increasingly important in the 21st century and if the polymer
industries are to retain their position in packaging technology they will
have to accept the challenge to modify their products and processes to

meet environmental demands. This will require the knowledge, skills and enthusiasm of the coming generation of young scientists who will hopefully be stimulated to accept this challenge.

I wish to express my thanks to my many collaborators and former students who have contributed to the understanding of polymer degradation and stabilisation and in particular to the late Professor Dan Gilead of Plastor Israel, to Dr. Sahar Al-Malaika of Aston University and Dr. Khirud Chakraborty of Robinson Brother Ltd. I am also grateful for helpful information provided by the following: Professor Ann-Christine Albertsson and Professor Sigbritt Karlsson of the Royal Institute of Technology, Stockholm; Dr. D. Biswas, Central Pollution Board, India; Mr. Jim Dodds of AMBRACO; Professor Emo Chiellini, Pisa University, Italy; Professor Fusako Kawai, Okayama University; Dr. Gerald Griffin, Epron Industries; Professor James Guillet, University of Toronto; Mr. Philip Law, British Plastics Federation; Professor Jacques Lemaire, Université Blaise Pascal, Clermont-Ferrand; Mr. Bert Lemmes, ORCA; Dr. Fred Mader, Association of Plastics Manufacturers in Europe; Professor Ramani Narayan, Michigan Biotechnology Institute, USA; Mr. D. A. Notman, BICC Cables Ltd.; Mr. Keith Richardson, Greater Manchester Waste Disposal Authority; Mr. Masaru Shibata, Chisso-Asahi Fertilizer Company; Dr. Graham Swift, Rohm & Haas Co.; and Dr. Irene Tan, Institute of Post-graduate Studies, Malaysia.

I am particularly indebted to my son, Ian Scott, Chair of Community Recycling Network, for his helpful comments on Chapter 4.

Gerald Scott

Contents

Chapter 1

Polymers in Modern Life

WHAT ARE POLYMERS?

Polymers are natural or man-made molecules, frequently called *macro-molecules*. They are composed of smaller units, *monomers*, which have reacted together to give a long chain, rather like a string of beads. In the simplest polymers, the monomers are identical and the polymer is named by prefixing 'poly' to the name of the monomer from which it is derived. Thus the polymer from ethylene is poly(ethylene), although in common usage the brackets are omitted. The monomers that constitute a polymer may be the same in which case they are called *homopolymers* or they may contain more than one monomer in which case they are *copolymers*. Additional monomers in a polymer may be randomly copolymerised to give *random copolymers* or may be polymerised in alternating blocks of identical monomers forming *block copolymers*.

–M–M–M–M–   M–N–N–M–N–M–   –M–M–M–N–N–N–
Homopolymer   *Random copolymer*   *Block copolymer*

Some polymers contain chemical bonds or *cross-links* between the long chains. Cross-links may be introduced during the synthesis of the polymer as is the case in the *thermosetting* polymers, which include the well known phenol–formaldehyde resins, but they may also be introduced into an existing polymer by a chemical reaction. This method of making cross-linked polymers is used in the process of rubber *vulcanisation* or sulfur cross-linking which was one of the earliest chemical reactions carried out on a naturally occurring polymer [*cis*-poly(isoprene)] obtained from the latex of the tree *Hevea braziliensis*. Many other polymeric materials are found in living organisms. The most common are the *polysaccharides*,

1

which include starch, the food store of seeds, and cellulose, the structural material of plants. A second important group are the polymerisation products of amino acids, the *polypeptides*, of which the proteins are widely distributed in living organisms. In the polypeptides the sequence of monomer units is much more complex than in the case of man-made polymers and the order in which the monomers are put together changes their nature and biological function.

$$-A_1-A_2-A_3-A_4-A_5-$$ where A_1, A_2, *etc.* are different amino acids
Polypeptide

Thus muscle, collagen (in bone), keratin (in hair, nails and beaks) and albumin are all copolymers of very similar amino acids but have quite different physical properties. In deoxyribonucleic acid (DNA), the genetic template, the sequence of monomers is precise and variations are the cause of genetic mutations. Although the polypeptides are of ultimate importance in life processes they are not important in the context of materials and will not be considered further in this book. However, they have had a significant impact on modern polymer science since the synthesis of the first man-made polyamide fibre, *Nylon*, by Carothers was modelled on the structure of a silk, a naturally occurring polypeptide.

NATURAL POLYMERS

Cellulose

The most easily recognised natural polymer is cellulose, the most abundant organic polymer on earth. It consists of glucose units and is the major component of wood although it is also found in the stems and leaves of many plants. Cotton is a particularly pure form of cellulose.

Cellulose

In woody materials, the long crystalline fibrils of cellulose are bound into a composite structure by lignin, a macromolecule based on polyphenols. Lignin, which is present to the extent of 25–30% in most woods, is a cross-linked polymer rather similar to man-made phenol–formaldehyde resins and may be looked upon as a 'glue' which gives wood its permanent form (Figure 1.1).

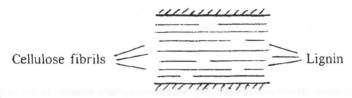

Cellulose fibrils   Lignin

Figure 1.1 *Structure of wood*

The overall effect is a very strong material that can bear enormous tensions and bending stresses without breaking. However, as every cricketer knows, wood can break under violent impact. This is because it is weakest along the direction of the fibres and the lignin which is a weaker polymer *delaminates* (separates) between the cellulose fibres. It is an interesting tribute to the evolution of natural materials that since the discovery of the synthetic polymers, man has employed the same principle of orientated fibre (glass or carbon) reinforcement in the manufacture of polymer composites with a strength (in the fibre direction) similar to that of steel.

Pure cellulose biodegrades relatively rapidly in the natural environment. Nature's abundant cellulosic litter in the form of leaves, grass, plant stems, *etc.* is bioassimilated in one season to give useful biomass. Branches and tree trunks take much longer to biodegrade and it is not always appreciated that, in some parts of the world, tree trunks and branches on the sea-shore present a much more significant litter problem than most of the commodity plastics. It may take decades and in some cases centuries for some fallen trees to disintegrate and biodegrade in the natural environment whereas a polyethylene container in the same situation would disappear due to photo-biodegradation in as many years. The situation may be quite different inside a forest environment, which is much more conducive to biological attack. Both cellulose and lignin biodegrade; the former much more rapidly than the latter by enzyme-catalysed *hydrolytic depolymerisation* of cellulose to its constituent sugars, which are assimilated by the cell. This process begins with the attachment of microflora to the *hydrophilic* (water-loving) surface of the tree trunk and the death of the microorganisms in turn provides humus for the growth of seedlings which eventually cannibalise the dead trunk. Polyethylene which is a *hydrophobic* polymer (that is it repels water) cannot undergo biodegradation unless it is modified by abiotic peroxidation. Sunlight catalyses this process and the rate-controlling process in the natural environment is photooxidation. This will be discussed in more detail in Chapter 5.

Wood-pulp cellulose is the basis of paper manufacture. The process involves separating the cellulose fibre from the resinous component of wood by treatment with alkali and carbon disulfide, an environmentally polluting process (Chapters 2, 5). Papermaking has been strongly criticised in recent years by ecologists due to the rapid depletion of the forests and this has resulted in an increase in recycling of used paper.

Cellulose fibres are crystalline and very strong materials when they are dry. However, they are hydrophilic and in the presence of moisture they absorb water, becoming permeable by microorganisms. For this reason paper became much less important as a food packaging material when the cheap hydrophobic synthetic polymers emerged in the second half of the 20th century.

The hydrophilic nature of cellulose is due to the high concentration of hydroxyl groups in the molecule. In the absence of water this gives rise to the association of the long molecules by *hydrogen bonding*. These 'cross-links' are weak compared to the valency bonds that hold together the repeating units in a polymer but because of the very large number of hydroxyl groups in cellulose, the molecules are held firmly together when dry to give the strong crystalline structure referred to above. Hydrophilic solvents can 'break' hydrogen bonds, resulting in swelling of the fibres. By using appropriate solvents, cellulose can be reformed into fibres (rayon) and films (cellophane) which retain the essential structure (and biodegradability) of cellulose. Replacement of the hydroxyl hydrogen in cellulose by hydrophobic groups such as methyl, ethyl or acetyl decreases its hydrophilicity. It can then be *plasticised* to give a thermoplastic material that can be processed to films. However, cellulose derivatives are more susceptible to water swelling than the hydrocarbon polymers, which are discussed below and they have been largely replaced by synthetic polymers in packaging applications.

Starch

Starch is the main energy storage system in plants and is closely related chemically to glycogen, the energy storage 'fuel' in animals. Starch consists of two polysaccharide components: amylose, a long unbranched chain of D-glucose units, and amylopectin, which is made from the same monomer units but cross-linked to give a much higher number of units in each molecule. It is, like wood, a composite structure in which the amylopectin provides the 'glue' that holds the starch granule together. Starch itself, unlike wood or paper, has no application as an industrial material but it will be seen in a later chapter that corn or potato starches have considerable potential as biodegradable thermoplastic polymers if appropriately plasticised. However, starch is much more readily hy-drolysed to glucose in the presence of microorganisms than cellulose and the main problem is to retain its integrity during use.

Wool, Hair and Silk

Like feathers and hair, wool consists mainly of keratin. This polypeptide contains L-cystine which is the oxidised (disulfide) form of L-cysteine. The former acts as a cross-link which determines the crimped nature of wool. In human hair this process is, of course, carried out reversibly in 'permanent waving':

$$—NHCCHCH_2SH \xrightarrow[\text{reduction}]{\text{oxidation}} —NHCCHCH_2S—SCH_2CHCNH— \quad (1.1)$$

L-Cysteine unit   L-Cystine unit

Silk, by contrast, consists of fibroin which does not contain cystine and is a straight fibre.

Natural Rubber [*cis*-poly(isoprene)]

The development of natural rubber as the first major industrial polymer was a very significant achievement of the 19th century. Rubber latex, as it comes from the tree, is a very unpromising material. It very rapidly loses its most useful mechanical property, namely elasticity, owing to attack by oxygen in the atmosphere (Chapter 3) which also transforms it into a very biodegradable material. Equally important to the development of rubber as an industrial product was the discovery of 'vulcanisation' (Vulcan, the God of fire) by Charles Goodyear in the USA and Thomas Hancock in England in 1839. Sulfur vulcanisation, or 'curing' as it is still called by rubber technologists, transformed the essentially thermoplastic polymer, which had poor dimensional stability, to a strong cross-linked matrix. Vulcanised rubber was found by Hancock to be a suitable material for bicycle tyres and in the 20th century it was developed for motor car tyres. A comparison of the environmental durability of an early bicycle tyre with that of a modern automobile tyre would be instructive but unfortunately none of the original rubbers now exist since they have all degraded to resinous non-resilient oxidation products. A modern tyre, unlike the early bicycle tyres, has to withstand the stresses and strains of high speed motorway travel (fatigue) at temperatures up to 100 °C and this is achieved by a complex combination of antioxidants which also give it protection against environmental pollutants such as ozone. Evidence for the environmental stability of modern tyre rubbers can be seen in many farmyards where discarded tyres are used as weights to hold down plastic covers. Some of these have been there for decades and show little sign of deterioration or biodegradation. This has nothing to do with the inherent biostability of the rubber molecule since we know that rubber latex peroxidises and biodegrades very rapidly when exposed to the outdoor environment. The durability of automotive tyres is rather a function of the exceptional effectiveness of the synthetic antioxidants that have been developed to protect the rubber from peroxidation during use. As will be seen later, the control of peroxidation is one of the mechanisms that is currently used in the development of polymers that can be programmed to biodegrade in a controlled way in the environment.

SYNTHETIC POLYMERS

Synthetic polymers are made by two main processes. The carbon-chain polymers result from the 'opening' of the double bond in the original olefin by *addition polymerisation* to give new carbon–carbon bonds:

$$n\text{RCH}{=}\text{CH}_2 \longrightarrow -(\underset{\underset{\text{R}}{|}}{\text{CH}}-\text{CH})_n- \qquad \text{Addition polymer} \qquad (1.2)$$

The hetero-chain polymers are mainly made by elimination of water between a carboxylic acid and an alcohol or an amine to give a polyester or a polyamide respectively. These are the *condensation polymers*:

$$n\text{HOC}\overset{\text{O}}{\overset{||}{\text{C}}}\text{R}\overset{\text{O}}{\overset{||}{\text{C}}}\text{OH} + n\text{HOR'OH} \longrightarrow \left[-\text{OC}\overset{\text{O}}{\overset{||}{\text{R}}}\overset{\text{O}}{\overset{||}{\text{C}}}\text{R'}-\right]_n + n\text{H}_2\text{O} \qquad (1.3)$$
$$\text{Polyester}$$

$$n\text{HOC}\overset{\text{O}}{\overset{||}{\text{C}}}\text{R}\overset{\text{O}}{\overset{||}{\text{C}}}\text{OH} + n\text{H}_2\text{NR'NH}_2 \longrightarrow \left[-\text{NHC}\overset{\text{O}}{\overset{||}{\text{R}}}\overset{\text{O}}{\overset{||}{\text{C}}}\text{NHR'}-\right]_n + n\text{H}_2\text{O} \qquad (1.4)$$
$$\text{Polyimide}$$

Similar products can be made without the elimination of water by 'ring-opening' polymerisation of monomeric cyclic monomers, the lactones and lactams. In this case the 'repeat unit' in the polymer is the same as that in the monomer:

$$\left[-\text{NH(CH}_2)_5\overset{\text{O}}{\overset{||}{\text{C}}}-\right]_n$$
$$\text{Polycaprolactam}$$

Polyolefins

This is the most important group of carbon-chain polymers made by the general reaction (1.2) in which R is hydrogen (polyethylene) or an alkyl group (*e.g.* CH_3 in polypropylene). In practice, polyethylene is not a single polymer. Three main sub-divisions are recognised.

Low Density Polyethylene

Low density polyethylene (LDPE) was the earliest polyolefin to be manufactured by the free radical polymerisation of ethylene. It is not 'pure' poly(methylene). The latter is a very linear polymer and can be synthesised in the laboratory by decomposition of diazomethane. As a result of its method of manufacture, LDPE contains alkyl groups of varying lengths pendant to the polymer chain and some carbon–carbon double bonds. These irregularities reduce its crystallinity compared with poly(methylene) so that it normally consists of 50–60% of crystalline domains dispersed in amorphous polymer. The amorphous matrix is a very viscous liquid which acts like a rubber and gives the polymer its very high resistance to impact. LDPE is still widely used because of its exceptional toughness.

High Density Polyethylene

High density polyethylene (HDPE) is made by a different (ionic) process which gives higher relative molar masses (M_r). This, coupled with its lower chain-branching (it is sometimes called 'linear' polyethylene), increases the ratio of crystalline to amorphous polymer which makes it much stronger than LDPE. However, the reduction in energy-absorbing amorphous phase also reduces its toughness.

Linear Low Density Polyethylenes

A compromise between LDPE and HDPE has been achieved in the commercial development of linear low density polyethylenes (LLDPEs), which are a series of polymers in which ethylene is copolymerised with a small proportion of other alkylethylenes. This gives controlled branching without sacrificing M_r or toughness.

Polypropylene

Polypropylene (PP) is made by a similar ionic reaction to HDPE and LLDPE, using organometallic catalysts. Unlike the ethylene polymers it exists in stereoisomeric forms due to the asymmetry of the methyl substituted carbon atom. The common industrial material is the *isotactic* form which can be stretched to give highly crystalline films and fibres. It is less tough than the LDPE but this may be improved by copolymerisation with a small amount of ethylene.

The major application of the polyolefins is in packaging (films and bottles). HDPE is aesthetically as well as technically an alternative to paper since it has the 'crinkly' feel of paper but unlike paper it has high resistance to water and hence microorganisms. PP is also widely used as a hydrophobic outer wrapping in the packaging of many commodities (*e.g.* cigarettes) where it has largely replaced cellophane, and in bottles as a replacement for glass.

There is little doubt that the emergence of the polyolefins in the mid-years of the present century transformed food distribution because of their effectiveness as sterile barriers against microbial contaminants. These advantages over glass, coupled with higher impact resistance and lower cost, were immediately obvious to the consumer but more recently it has also been recognised that the use of plastics in the packaging of foodstuffs is much less energy intensive than the use of traditional packaging materials (Chapter 2).

An important use of polyethylene films is in 'shrink-wrap' packaging in which the films are given a 'built-in' stress which is released on heating with contraction of the film over goods to be packaged. This is a convenient and economic alternative to traditional 'boxing' in cardboard. A rather similar packaging application is 'stretch-wrap' polyethylene film for hay and silage. This mainly black wrapping film has replaced hay stacks in the countryside and is also a major source of plastics pollution in the countryside (Chapter 5).

Another important use for the crystalline polyolefins is in industrial and, to a lesser extent, domestic fibres. Polypropylene fibres have also largely replaced the naturally derived sisal in agriculture. This fibrous material physically resembles sisal but unlike the natural product it does not readily biodegrade during use or after discard. Agricultural twines are often made very cheaply by slitting uniaxial stretched polypropylene sheets in the direction of the applied stress. The non-biodegradability of discarded PP twines has been overcome in recent years by making the polymer photo-biodegradable (Chapter 5).

Plastics have achieved a dominant position in agriculture and horticulture over the past twenty years, which has merited the new description *plasticulture*. This is in part due to the replacement of glass in greenhouses and tunnels by polyethylene, but more particularly in the development of polyethylene '*mulch*'. Polyethylene for greenhouse films have to be as stable as possible to the environment and in spite of the development of synergistic 'light stabilisers' (Chapter 3) the maximum lifetime that can be currently achieved even in the less sunny climes is about three years. After this time slits appear in the films from stress points, particularly where the films are in contact with the greenhouse structure and the material has to be replaced. In spite of this it is increasingly more economical to use polyethylene than glass due to the lower cost and shatter resistance of the former.

Mulching films, unlike greenhouse films, are generally dark pigmented and are laid on the surface of the soil where they not only increase the temperature of the soil and reduce weed growth but they also reduce the usage of irrigation water and fertilisers contained in them. This specialised use of the polyolefins in agricultural systems will be discussed in more detail in Chapters 2 and 5.

Ethylene–Propylene Copolymers

When ethylene and propylene are randomly copolymerised with a propylene content of between 30 and 60 mol % the polymers are no longer semi-crystalline but behave like typical rubbers. Unfortunately, ethylene–propylene rubbers (EPR) are more difficult to cross-link in the conventional way with sulfur although, like polyethylene, they can be cross-linked with dialkyl peroxides to give rubbers that are much more resistant to environmental peroxidation and degradation than the poly(diene) rubbers (see below). The EP rubbers can be made cross-linkable by sulfur by the introduction of a low concentration of a diene (generally 5–15 double bonds per 1000 carbon atoms). These copolymers, normally referred to as ethylene–propylene–diene terpolymers (EPDM), are not as resistant to oxidation as EP itself, but for many applications this can be compensated for by adjusting the antioxidant system. For very demanding applications (*e.g.* in seals and gaskets) peroxide cross-linking is used together with antioxidants that resist leaching from the polymer. This will be discussed in Chapter 3.

Polydienes

The poly(1,3-dienes) are made by similar chemical processes to the polyolefins:

$$n\text{CH}_2{=}\text{CHC}{=}\text{CH}_2 \longrightarrow [-\text{CH}_2\text{CH}{=}\overset{\text{R}}{\text{C}}\text{CH}_2-]_n \qquad \text{Poly(1,3-dienes)} \qquad (1.5)$$

Natural rubber (pages 5 and 6) is a member of this group of polymers although it is synthesised in the rubber tree by a quite different process. NR is a high M_r member of the terpene family and is made by the same kind of enzyme-controlled chemistry as the terpene squalene, which is the precursor of cholesterol in animals. Squalene is often used as a low M_r 'model' to study the reactions of *cis*-poly(isoprene) (PI).

$$\left[\begin{array}{c} H_3C \\ \diagdown \\ CH=CH \\ H_3C \diagup \end{array} \begin{array}{c} CH_2CH_2 \\ \diagdown \\ C=CH \\ H_3C \diagup \end{array} \begin{array}{c} CH_2CH_2 \\ \diagdown \\ C=CH \\ H_3C \diagup \end{array} \begin{array}{c} CH_2 - \\ \diagdown \\ C=CH \end{array} \right]_2$$

Squalene

The earliest rubber to be manufactured synthetically was not polyisoprene but the copolymer of butadiene and styrene (SBR) by random free radical polymerisation. Modern SBR contains styrene and butadiene units in the ratio 1:3.

The structure of *cis*-polybutadiene (*cis*-PB) is more complex than that of *cis*-PI since some of the double bonds react by 1,2 addition to give a pendant vinyl group which is more reactive toward radical addition than the vinylene group in *cis*-PI.

$$-CH_2CHCH_2CH=CHCH_2CH_2CH-$$
$$\underset{CH=CH_2}{|}$$

Polybutadiene rubber (*cis*-PBR)

The same ex-chain double bond structure is also present in SBR and leads to cross-linking through oxygen during ageing.

More recently, block copolymers of styrene and butadiene (SBS) have been made using organometallic catalysts and they have been found to have valuable additional properties compared with the random copolymers of SBR. They have been given the generic name *thermoplastic elastomers*, TPEs, because, although they can be processed like SBR, on cooling they have much better dimensional stability at ambient temperatures and show some of the properties of cross-linked rubbers. This results from the association and phase separation of the polystyrene segments to give glassy domains similar to those existing in polystyrene itself (see below). Since the chemically attached polybutadiene segments are excluded from the PS domains they retain their elasticity and the PS domains behave as macromolecular cross-links (Figure 1.2).

Neither SBR nor *cis*-PB contain any appreciable amount of crystallinity and in this respect they differ from natural rubber, which crystallises slowly at ambient temperatures. Crystallisation is a disadvantage before vulcanisation but is a major advantage when subjected to stress and strain in a tyre. The higher the applied stress, the greater is the crystallinity and

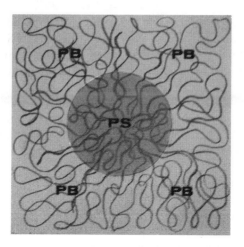

Figure 1.2   *Structure of styrene–butadiene block copolymer (SBS)*

hence strength of *cis*-PI. Unlike the orientation of PP referred to above, this is a reversible process and when the stress is removed it returns to its normal non-crystalline resilient state. SBR is the major rubber component of motor car tyres but *cis*-PI (NR) is preferred in truck tyres since it generates less heat during high speed running.

Quite different polymers can be obtained from butadiene by polymerisation through the 1,2 position. 1,2-PB is no longer a rubber but a semi-crystalline thermoplastic polymer rather similar to PP which has some potential as an environmentally degradable plastic (Chapter 5).

Chlorine-containing rubbers based on the polymerisation of chloroprene [reaction (1.4), R = Cl] are speciality polymers with good resistance to organic solvents and chemical reagents. Nitrile–butadiene rubber (NBR) has similar properties and is obtained by randomly copolymerising acrylonitrile with butadiene:

$$\left[-CH_2\overset{\displaystyle CN}{\underset{|}{C}}HCH_2CH{=}CHCH_2- \right]_n$$

Nitrile—butadiene rubber (NBR)

Polystyrene

Polystyrene (PS) is a brittle transparent (non-crystalline) polymer with a glass–rubber transition temperature (T_g) just below 100 °C. Like silica glass it has a high intrinsic tensile strength but relatively low impact resistance compared with the polyolefins. For this reason, the use of unmodified ('crystal') polystyrene is limited but it can be readily modified by grafting styrene monomer onto a polybutadiene backbone (5–15% PBD) to give 'high-impact' polystyrene (HIPS). This is much less transparent than unmodified PS but its impact resistance is increased up to four times. Unfortunately, the incorporation of a rubber phase decreases the exceptionally good light resistance of PS by increasing its photo-peroxidisability. Consequently it discolours and embrittles much more rapidly than pure polystyrene and, for long-term durability, HIPS requires substantial amounts of antioxidants and stabilisers to give it the necessary resistance to peroxidation (Chapter 3).

A major use for polystyrene is in expanded (bead) form. This material is relatively tough and is used mainly as a protective packing material, particularly for electrical equipment or as sintered and shaped 'boards' to replace paper, cardboard, *etc.* in packaging containers. Like the polyolefin packaging materials, PS is often observed as a non-degradable environmental pollutant on the sea-shore due to 'escape' of beads and pieces of broken PS foam packaging.

Styrene is also used as a basis of copolymers with other monomers. Styrene–acrylonitrile copolymer (SAN) has properties rather similar to PS but is somewhat tougher. Acrylonitrile–butadiene–styrene (ABS) copolymers, on the other hand resemble HIPS and are manufactured by a similar graft-copolymerisation process. This material has proved to be very useful for computer housing but like HIPS it is not very environmentally stable and discolours readily.

Poly(vinyl chloride)

Poly(vinyl chloride) (PVC) is essentially a glassy (non-crystalline) polymer like PS. It is, however, tougher and when pigmented it can be used out-of-doors in exterior cladding, window frames, *etc.* where it is commonly designated *UPVC* ('unplasticised PVC'). This is a remarkably durable material in view of the known thermal- and photo-instability of PVC and this is achieved by the addition of substantial quantities of stabilisers (Chapter 3). The present trend is to incorporate light stable impact modifiers based on saturated polyacrylates to increase toughness in the outdoor environment.

PVC degrades rapidly in heat and light unless it is effectively stabilised. The most obvious effect of degradation is discolouration; initially yellowing but ultimately blackening. It is widely used in rooflights and the formation of brown discolouration is the initial indication that the useful properties of the plastic (particularly impact resistance) are also deteriorating. Nevertheless, even transparent PVC can be made relatively durable to the outdoor environment by the use of organometallic stabilisers (Chapter 3). Rigid PVC is also used as a replacement for glass in the bottling of mineral waters and even of wine. In this form it ends up in substantial quantities in the domestic waste stream. Commercial PVC has been the subject of much criticism in the false belief that, like some low molecular weight chlorine compounds, it is intrinsically toxic. This is not so. The polymer is chemically and biologically inert unless it is degraded, when hydrogen chloride and lower molecular weight chlorine compounds are formed. The use of PVC in packaging is controversial because of the potential toxicity of chlorine compounds formed from it during incineration. Over 50% of its weight is chlorine which is liberated as HCl and this has to be neutralised during incineration. Furthermore the residues from the organometallic stabiliser may be toxic and have to be disposed of carefully. The problem of PVC disposal will be discussed in more detail in Chapter 4 where alternative stratagems for dealing with plastics waste will be discussed but it should be noted here that of the alternatives available, only mechanical recycling appears to offer the prospect of minimising the chlorine hazard.

A traditional use of PVC is in 'leathercloth'. The flexibility of PVC in this kind of application is achieved by the use of plasticisers which are generally present from 30–35% in a leathercloth. Plasticisers, which are normally aliphatic esters of aromatic or aliphatic dicarboxylic acids are essentially solvents for the polymer and they have the practical effect of reducing the glass–rubber transition temperature (T_g) so that at ambient temperatures the material behaves more like a rubber than a plastic. Unfortunately, plasticisers, like the impact modifiers discussed above, peroxidise more readily than the PVC molecule, particularly in light, and the peroxides produced act as sensitisers for the photooxidative deterioration of leathercloth. This is why furniture covered with PVC leathercloth frequently shows evidence of embrittlement and cracking on surfaces exposed to sunlight even through glass windows. Leathercloth presents a similar problem in the ultimate disposal of furniture but this is of a lower order than the problem of disposing of PVC packaging.

Poly(vinyl esters) and Poly(vinyl alcohol)

Poly(vinyl acetate) (PVA) has a glass–rubber transition temperature in the region of room temperature and is therefore neither a useful rubber nor a useful plastic. It can, however, be plasticised like PVC to give a rubbery material with limited practical utility. Copolymerisation of vinyl acetate with ethylene in relatively low ($\sim$ 3 mol %) concentration (EVA) is a useful way of introducing polar properties into hydrocarbon polymers. EVA has found a very important application as 'cling film' due to its ability to form a good seal with ceramic or metal surfaces. Higher concentrations of VA give ethylene copolymers with properties rather similar to plasticised PVC. However, unlike plasticised PVC, from which the phthalate plasticisers can be removed from packaging by fats and oils present in foodstuffs in contact with it, no such leaching can occur from EVA and the latter has now largely replaced PVC films in the packaging of foodstuffs. An additional advantage of course is that during disposal it does not generate chlorine compounds on incineration.

Poly(vinyl alcohol) is made from poly(vinyl acetate) by hydrolysis. This polymer is water soluble and, as will be seen in Chapter 5, is one of the few synthetic water-soluble polymers to undergo rapid biodegradation in sewage systems.

Poly(acrylates), Poly(methacrylates) and Poly(acrylic acids)

Poly(methyl methacrylate) (PMMA) is the well known tough transparent polymer Perspex. It is one of the most light stable polymers known. Aircraft windows dating from the early 1940s which have been continuously exposed to the environment since that time have remained virtually unchanged apart from surface crazing. Unlike most thermoplastic polymers, PMMA is not produced by extrusion but by dissolving in a solvent (often the monomer) and casting into sheets. Monomer is removed by further polymerisation. It will be seen later that processing in an internal mixer is a damaging process and the great light stability of PMMA is due, at least in part, to the method of manufacture.

The higher homologues of the poly(alkyl acrylates) are rubbery materials and are sometimes used alone or as copolymers as impact modifiers for glassy polymers such as PS or PVC. Poly(acrylic acid) does not behave as a thermoplastic polymer due to its strong hydrogen bonding. It is water soluble and when acrylic acid is graft copolymerised with the polyolefins, the carboxylic acid groups profoundly modify these hydrophobic polymers, improving their ability to be reinforced by polar fibres such as glass.

Polyesters

Poly(ethyleneterephthalate) is the fibre-forming polymer commonly known as PET.

Poly(ethyleneterephthalate) (PET)

As manufactured, PET is a relatively weak, amorphous (non-crystalline) polymer which has to be crystallised and orientated by stretching to give it its textile properties. Orientation is carried out during the 'spinning' operation. Equally important nowadays is the use of PET as a replacement for glass in soft drink bottles. Although more expensive than the polyolefins or PVC, it has a distinctive appearance and can be readily separated from a waste stream. Its recovery for recycling will be discussed in Chapter 4.

The polyester (alkyd) resins based on phthalic anhydride and glycerol have been used as surface coatings for many years, but they are also the basis of a large number of cross-linkable polymers used in engineering materials. These generally contain a proportion of an unsaturated dicarboxylic acid derived from maleic or fumaric acids:

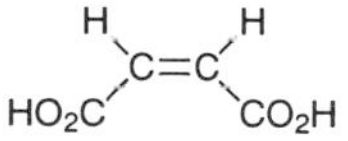

which can be cross-linked by a free radical process to give tough composite materials when reinforced with glass fibre (generally called 'fibre-glass'). Artifacts made from glass reinforced polyesters are, like PET, very resistant to the environment but very rarely appear in the domestic waste stream.

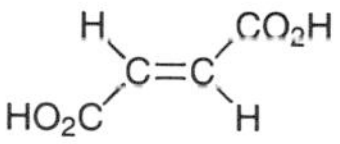

Polyamides

The most important polyamides are the fibre-forming nylons, notably Nylon 66 and Nylon 6:

$$[-CO(CH_2)_4CONH(CH_2)_6NH-]_n \qquad\qquad [-CO(CH_2)_5NH-]_n$$

Nylon 66   Nylon 6

These polymers, like the polyesters, contain hydrogen bonds between the polymer chains and when 'drawn' into fibres they become highly crystalline and hence strong in the direction of the extension, which makes them suitable not only for domestic fibres and textiles but for many industrial applications (*e.g.* ropes, tyre cords, *etc.*). The polyamides are relatively light-stable polymers which makes them particularly suitable for outdoor use. They do, however, yellow in sunlight over a period of time due to peroxidation and are normally pigmented to minimise this effect. Other nylons are derived from longer chain dicarboxylic acids [reaction (1.3), $R = C_{10}H_{20}$] or lactones. Some of these have interesting properties as 'engineering plastics', for example in gear wheels, where their self-lubricating properties lead to good wear resistance in contact with metals.

Thermosetting Resins

One of the oldest group of polymers, still widely used in the electrical industry as 'Bakelite', is the phenol–formaldehyde (PF) resins formed *in situ* by reaction of phenol with formaldehyde. Owing to the polyfunctionality of phenol which can react with three difunctional formaldehyde molecules, the result is a highly cross-linked structure which can be formed in one step from the low M_r precursors.

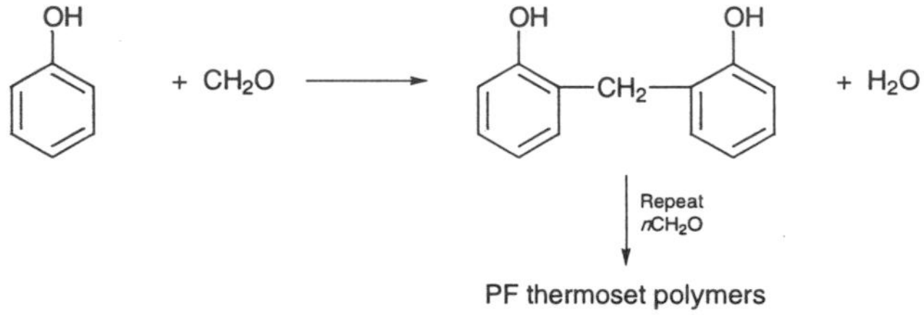

PF thermoset polymers

Similar cross-linked macromolecules can be prepared by reaction of formaldehyde with urea (UF), and by melamine (MF) with formaldehyde. These are hard but tough materials which are much less coloured than PF. They are the materials used in white electrical plugs and in pigmented form as 'Formica' for working surfaces.

The main environmental limitation of the cross-linked resins is that they cannot be readily mechanically recycled. They can of course be burned and may be ground to powder and used as inert fillers for the thermoplastic polymers. However, since they are intended to be durable materials, they represent no real threat to the environment.

Poly(urethanes)

Poly(urethanes) are the reaction products of di-isocyanates and diols. By varying the functionality of the diols and the addition of water, the products of the reaction vary from cross-linked foams to rubbers:

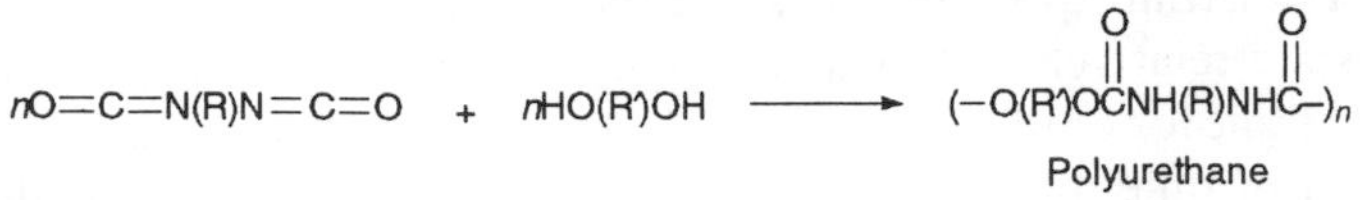

$$n\text{O}{=}\text{C}{=}\text{N(R)N}{=}\text{C}{=}\text{O} \ + \ n\text{HO(R')OH} \longrightarrow (-\text{O(R')OCNH(R)NHC-})_n$$

Polyurethane

In modern poly(urethane) technology, the simple diols such as ethylene glycol which were originally used have been replaced by hydroxyl-ended polyesters and polyethers. The properties of the poly(urethanes) then depend on the nature of the polyesters and polyethers as much as on the di-isocyanate. Typical examples of hydroxyl-ended polyesters and polyethers are poly(ethylene adipate) (PEA) and poly(propylene glycol) (PPG):

$$HO[(CH_2)_2OC(CH_2)_4CO]_n(CH_2)_2OH$$

PEA

$$HOCHCH_2(OCHCH_2)_nOH$$

PPG

Poly(urethane) foams based on polyethers have now largely replaced polydiene rubbers in upholstery and flammability is a major disadvantage compared with traditional upholstery. A major problem is that it is not the fire itself that kills people but the toxic fumes that are produced in the smoke and this is exacerbated by certain types of flame retardant. There are no simple solutions to this problem. Foams in their very nature have a large surface area and a developing fire thrives on the accessibility of 'fuel' from the exposed foam (Chapter 3). The most promising solution is to make the textile fabric surrounding the foam non-flammable so that the fire never reaches the foam itself.

Poly(urethanes), like the polyamides, are attractive to rodents who use them as a source of nutrients because of their nitrogen content. Poly(urethane) foams are also micro-biodegradable if left in humid environments and the loss of strength due to biodegradation is a significant problem in some applications.

Epoxy Resins

The epoxy resins are made by reaction of a diol (*e.g.* bisphenol A) with an epoxide, of which epichlorohydrin is typical. Since both the chlorine and the epoxy group in epichlorohydrin are reactive toward the hydroxyl group in bisphenol A, a linear polymer is produced:

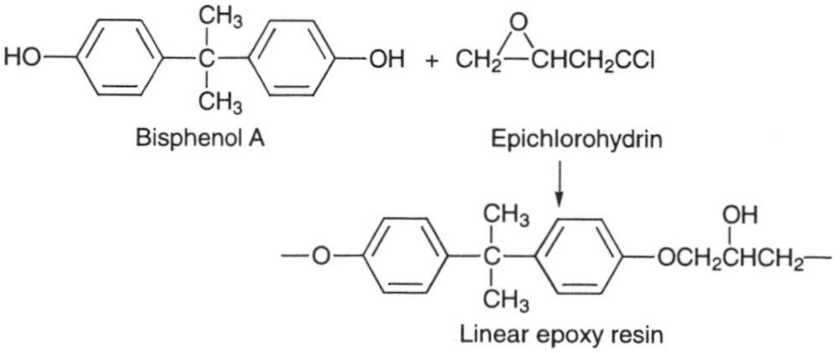

Bisphenol A   Epichlorohydrin

$$—OCH_2CHCH_2—$$ (1.6)

Linear epoxy resin

Epoxys can be cross-linked by co-reacting an amine or an acid anhydride to give tough, dimensionally stable products. In recent years these have been the basis of new engineering composites when reinforced with glass or carbon fibres.

Epoxy resin composites, because of their specialised engineering uses, have no significant deleterious effects on the environment after use. They do, however, make a major contribution to energy saving in ship hull and aircraft construction. Polymeric components are generally much lighter than their equivalent metal counterparts and it will be seen in Chapter 2 that plastics involve less energy input during their production than metals. This leads to saving energy both during manufacture and in use and it is inevitable that, as the mechanical performance of polymer composites is improved still further, they will play an increasingly important role as materials of construction.

Speciality Polymers

A variety of polymers are used in engineering and medical applications which have relatively little impact on the environment. These are mainly high performance and relatively expensive polymers such as the *silicones* in rubbers, the specialised polyamides referred to above in gear wheels, *polycarbonates* (in office equipment), chlorinated and sulfonated rubbers, fluorinated polymers such as *poly(tetrafluoroethylene)* (*Teflon*™) in metal coating and the *polyimides* which, owing to their 'ladder' structure, are extremely stable in high temperature applications. Since these polymers are high-cost durable materials, they rarely appear in the waste stream.

FURTHER READING

1 J. W. Nicholson, *The Chemistry of Polymers*, 2nd Edn, Royal Society of Chemistry, Cambridge, 1997.
2 K. J. Saunders, *Organic Polymer Chemistry*, Chapman & Hall, London, 1973.
3 J. A. Bridson, *Plastics Materials*, Newnes-Butterworths, 1979.
4 *Rubber Technology and Manufacture*, ed. C. M. Blow, Butterworth, 1971.
5 J. M. G. Cowie, *Polymers: Chemistry and Physics of Modern Materials*, Intertext Books, 1973.
6 *Advances in Polyolefins*, ed. R. B. Seymour and T. Cheng, Plenum Press, 1987.

Chapter 2

Environmental Impact of Polymers

WHY POLYMERS?

The earliest industrial polymers were derived from natural products, either by extraction from biological sources, as in the case of natural rubber (NR), or by conversion by chemical reaction from natural products, as in the case of paint films (see page 27). Both still play an essential part in modern technology even though their uses have been complemented by synthetic rubbers and surface coatings. A great deal of work has been done by the natural rubber producers to present NR in a similar form to the synthetic analogues and with similar technical specifications. At the present time natural *cis*-PI is still rather superior in some properties to the more recent synthetic analogue. Similarly paints based on the polyunsaturated fatty esters are still competitive with more recent synthetic products.

Perhaps the most remarkable aspect of polymers derived from natural products is that such environmentally unstable materials as NR and linseed oil should be the basis of environmentally durable industrial products. The history of the motor car tyre illustrates the technologist's ability to design products for durability. Tyre rubbers before fabrication are among the least environmentally stable of all polymers and yet motor car tyres survive for many years in the outdoor environment long after use. It was not always so. During the early years of this century, the environmental stress on the automobile tyre was very much less than it is in these days of high speed motorway travel. The rubbers used at that time would not have survived for a fraction of the time under conditions of modern travel. Over this period the durability of tyre rubbers has evolved with the development of heat stabilisers, antifatigue agents and antiozonants (Chapter 3).

Plastics in Packaging

In this century, the major use of synthetic polymers has been as replacements for more traditional materials, particularly in packaging. Today the packaging industry is by far the major user of plastics. Over 60% of post-consumer plastics waste is produced by households; most of it as single use packaging. Several factors have been responsible for the phenomenal growth of the use of plastics in packaging. The thermoplastic polymers are light in weight and yet have very good barrier properties against water and water-borne organisms. Compared with glass they have much superior impact resistance and resilience, resulting in reduced product losses during transport. They therefore protect perishable commodities from the environment on the one hand and protect the environment from corrosive or toxic chemicals on the other.

The production processes for plastics from crude oil to fabricated product are much less labour and energy intensive than traditional materials. Table 2.1 compares the energy usage in the manufacture of typical packaging materials.

Table 2.1 *Energy^a requirements for the production of materials used in packaging*

Material	Energy requirement/$kWh\,kg^{-1}$
Aluminium	74.1
Steel	13.9
Glass	7.9
Paper	7.1
Plastic	3.1

^{a}Energy utilisation is related to cost and to the CO_2 burden on the environment ('greenhouse effect').

The fabrication of plastics by injection moulding is also less energy intensive than the fabrication of traditional materials. The liquid polymer is converted into the final product in a single rapid and repetitive process that does away with intermediate forming and joining procedures. When these factors are combined with the lower density of polymers, the energy requirements for similar containers are found to be very much less than for traditional materials (Table 2.2). Even if no energy were involved in the transport and cleansing of returnable bottles (see below), returnable bottles would have to be recycled about twenty times to compete with plastics.

It is not always fully appreciated by environmental campaigners that the manufacture of paper is more environmentally polluting than the

Table 2.2 *Energy requirements for similar beverage containers*

Container	Energy usage per container/kWh
Aluminium can	3.00
Returnable soft drink bottle	2.40
Returnable glass beer bottle	2.00
Steel can	0.70
Paper milk carton	0.18
Plastic beverage container	0.11

manufacture of polymers. Table 2.3 compares the effluents emitted in the manufacture of 50 000 shopping bags from paper compared with similar bags from polyethylene. Sulfur dioxide and oxides of nitrogen are both damaging to human health and to trees and vegetation and their reduction is clearly a step towards a cleaner environment.

Another application of plastics in packaging is the black plastic sheeting widely used in the outdoor storage of hay. This material is very stable after discard and since it is carried long distances by the wind it represents a major pollution nuisance in the countryside (Figure 2.1). A similar problem is caused by the replacement of naturally occurring sisal by polypropylene twines in the baling of hay. Orientation of PP sheet allows the film to be slit in the direction of orientation into a strong fibrous material with mechanical properties as good as or better than natural fibres. However, unlike the natural products, PP twines do not biodegrade naturally and their use in agricultural packaging has led to a litter nuisance in the countryside which is also a danger to animals when they attempt to ingest it. This partly results from overstabilisation during manufacture and baler twines are now manufactured in the USA which photodegrade and biodegrade after they have fulfilled their primary function (Chapter 5).

Table 2.3 *Effluent emitted (kg) during the manufacture of 50 000 shopping bags*

	Polyethylene	Paper
Sulfur dioxide	10	28
Oxides of nitrogen	6	11
Hydrocarbons	3	2
Carbon monoxide	6	2
Dust	1	3

Figure 2.1   *Wind-blown stretch-wrap film*

Polymers are very light compared with metals and ceramics. The modern plastic milk container is only a fraction of the weight of a similar bottle made from glass and this has a very significant influence on transport costs. It is a popular belief that milk delivery in returnable glass bottles is ecologically preferable to singe-trip plastic containers. However, it has been shown that if a major city such as Munich were to return to traditional milk distribution in glass bottles, an additional 2700 milk floats would be required and an additional 240 000 litres of diesel fuel would be used daily. In addition, of course, the bottles would have to be cleansed daily with further energy consumption and waste production. A complete energy balance is required in every case before a choice can be made between traditional materials and 'cradle-to-grave' assessment (*life-cycle assessment, LCA,* Chapter 4) is now widely used to describe total environmental impact. LCA would therefore take into account not only the energy that goes into the production and use of the initial package but also into its ultimate disposal. It will be seen in Chapter 4 that once obvious methods of disposal of redundant materials, namely tipping or land-fill, are no longer environmentally acceptable options. Alternative stratagems, such as materials recycling, re-use, *etc.,* involve the input of some additional energy. However, plastics differ from metals, glass and ceramics in that the heat content of the material itself can be used to produce energy and this displaces the energy balance even further in favour of plastics (Table 2.4).

Table 2.4 *Energy consumed during manufacture and produced during disposal for plastics and glass in milk containers*

Container	Energy used in manufacture/kWh	Energy produced during combustion/ kcal
'Two quart' glass milk bottle	8.36	0
'Two quart' plastic (PE) pouch	0.84	317

Polymers in Transport

Plastics are also increasingly replacing traditional materials in automotive components, for example in motor vehicles, aircraft and boats. Several physical characteristics contribute to their increasing popularity but the most important is reduced weight. All the plastics materials have densities in the region of one, whereas even the lightest of the commonly used metals, aluminium, is considerably heavier. It has been estimated that over the past decade 200–300 kg of traditional materials have been replaced by 100 kg of plastics in the average car with the consequent reduction of fuel consumption by about 750 litres over the useful life of the car ($\sim 150\,000$ km). This has resulted in an estimated reduction in oil consumption of $\sim 12\,MT$ and CO_2 emissions by $\sim 30\,MT$ across Western Europe. In addition, since much more energy is used in the production of metals and glass than the manufacture of the common plastics, it follows that the more plastics that can be used to replace metals and glass in vehicles, the less fuel will be used in transport. Engineering plastics are increasingly used in structures such as decks in aircraft and in boat hulls, masts, *etc.*, very often in the form of fibre-reinforced composites. Their primary limitation is in high temperature applications, particularly in engines, owing to their relatively low melting/decomposition temperatures. Nevertheless, polymer composites are used in the heat absorbing shields of rockets, where porous carbonaceous decomposition products formed by *ablation* act as thermal insulation during re-entry to the Earth's atmosphere.

Rubbers have also become indispensable in automobiles due to their unique elasticity in tyres and shock absorbers. The tyre has itself undergone evolution over the years to give it durability against the environmental stresses experienced in high speed transport and improvements in 'rolling resistance' have reduced the energy used in cars and trucks.

Plastics in Agriculture

Plastics have changed the face of the rural environment. Although they comprise less than 2% of total plastics usage in Europe and about 4% in the USA, very much more is used in the Mediterranean countries (Spain, 8%, Israel, 12%) and in China (20%) where agriculture is much more intensive. Not only have plastics largely replaced glass in greenhouses and tunnels, but they have made possible the introduction of mulching films, a novel agricultural technology not possible before the introduction of man-made plastics (Figure 2.2). Biodegradable mulch from natural sources has been used since time immemorial to provide an insulating layer round the roots of vegetables and soft fruits. The word 'mulch' is derived from the Greek word for decaying, and straw has been used for centuries to protect strawberries from rotting or soiling (hence their name). Plastics have largely replaced these traditional materials and the fact that they do not decay rapidly in use is both an advantage and a disadvantage. It is an advantage because it ensures a coherent protective barrier between the roots of the plants and the outside environment throughout the growing life of the plant but it is a disadvantage after the crop has been harvested because the films have to be collected and disposed of manually at considerable cost to the farmer.

The use of plastics in modern agriculture has been given the name *plasticulture* because of its importance in the technology of large-scale food production. Plastic mulching films not only raise the temperature of the soil by several degrees, but they also reduce water loss, particularly in arid climates where water availability is the primary limitation to successful agriculture and horticulture. The production of one ton of grain requires about 1000 tons of water and the availability of water will be the main limitation in the production of food during the next century. Plastics films, owing to their excellent water barrier, not only reduce the usage of irrigation water to a fraction of that used in bare soil cultivation but the film also creates a microclimate for the roots of the plants, maintaining a more even temperature and a constant humidity by irrigation. Under these conditions the yields of most crops are substantially increased (Table 2.5).

It is imperative that the integrity of the film does not break down until the end of the growing season, otherwise yields are considerably reduced. There have been law suits in the USA in which millions of dollars in damages have been awarded to farmers against suppliers of mulching film cheapened by the incorporation of recycled polymer with resulting premature failure of the microclimate beneath the film. Nevertheless, mulching films should ideally disintegrate and subsequently biodegrade

(a)

(b)

Figure 2.2 *Polyethylene mulching film* (a) *during growing of strawberries*
(b) *removal of residues after cropping*

after cropping to permit automated cropping equipment to be used
without clogging the machinery. Technologies to achieve continuous
protection during crop growth followed by polymer biodegradation
after use will be discussed in Chapter 5.

Table 2.5 *Crop yields with and without mulching film*

Crop	Location/ treatment	Yield/ kg	Increase (%)
Musk melons	New Brunswick/Bare ground	62.73	
(per 7 plants)	Mulched	92.64	47
Tomatoes	Oregon/Bare ground	10.5	
(per plant)	Mulched	15.2	45
Bell pepper	Rio Grande/Bare ground	1356	
(per hectare)	Mulched (mid-bed trenching)	6633	389

Plastics are much cheaper than glass in greenhouses but they have to be replaced more frequently. Glass is normally destroyed in accidents, whereas plastics degrade fairly predictably due to the weather. Stabilisation is thus the key to the economics of the replacement of glass in greenhouses and this in turn depends on the development of polyolefins stable to the environment. Recently developed light stabilisers based on light-stable photo-antioxidants (Chapter 3) have moved the economic balance toward plastics. However, this is still a very active area of research directed towards further improvements in the durability of the polyethylenes *versus* that of glass. This requires a thorough investigation of all the factors involved since both heat and light determine the deterioration of polymers out-of-doors. The effect of both factors together is always greater than the sum of the individual effects. Furthermore, some pest-control chemicals used in greenhouse cultivation reduce the effectiveness of the commonly used light stabilisers, leading to reduced durability of greenhouse plastics. Failure of greenhouse films generally occurs where the film is in contact with the framework structure due to a combination of heat and light and the most effective stabilisers are light-stable antioxidants (Chapter 3, pages 59–61). Heat stabilisation is particularly important if the damaged films are recycled. The main bulk of the plastic remains unchanged and with careful reformulation the recovered plastic can be blended with virgin polymer in a 'closed loop' to the same or a similar application.

In addition to the large quantities of black plastic hay-wrap referred to above, which are discarded in the fields and farmyard, very large amounts of polyethylene and polypropylene are used in agricultural packaging of feedstuffs and fertiliser sacks. In spite of the fact that some of this is recycled and farmers find secondary uses for some of it, much of it finds its way into rivers and ditches where it degrades very slowly. This is the environmental down-side of plastics packaging.

Polymers in the Home and Office

It is now taken for granted that for equipment operating at ambient temperatures, plastics are the modern materials of choice for such items as food mixers, vacuum cleaners, hair driers, television consoles, computers, word processors and other office equipment. Most of these are produced by injection moulding in impact-resistant materials and can be produced accurately and rapidly. Furniture is also largely upholstered with polymeric materials. At one time furniture was hand-built around wooden frames with straw or horse-hair upholstery. Now the padding has been replaced by moulded foams. These are based either on rubber latices which may be foamed mechanically or by the use of nitrogen-generating foaming agents during vulcanisation. The poly(urethanes) (PUs) produce their own foaming agent (CO_2) as a side-reaction of the chemistry of iso-cyanate reactions with polyols in the presence of a small amount of water. PU foams in particular have a much lower bulk density than traditional upholstery. The surface finish of most furniture is either PVC 'leathercloth' (page 13), which simulates the much more expensive leather originally used in 'quality' furniture, or synthetic fabrics based on either polyester or polyamide fibres. Wood and metals are still used in the primary load-bearing components of beds, armchairs, settees, *etc.* and of course in 'quality' tables, chairs, bookshelves, *etc.* However, decorative wooden furniture is frequently given a polymeric coating [generally poly(urethane)] for protection. Plastics are also used more subtly in modern decorative reproductions such as 'ancient' beams. Artificial plastics flowers often look as 'real' as the natural product and normally last much longer!

Paints and Surface Coatings

Naturally occurring 'drying oils' based on polyunsaturated fatty esters (*e.g.* linoleates) have been used for centuries to protect metals from corrosion and wood from biological action. In the early 1970s more than 90% of paints and coatings sold worldwide were applied in organic solvents or solutions. As a result of the oil crisis of the mid-1970s, it was realised that the wastage of organic solvents with the associated energy costs could not be allowed to continue and the emphasis changed to reduce or eliminate organic solvents using lower energy drying or curing processes. Following the introduction of 'Clean-air' Acts, many plasticisers and solvents (*e.g.* methanol, methylene dichloride, methyl ethyl ketone, toluene, xylene, *etc.*) formerly permitted in surface coating technologies were classified as 'hazardous' and this led to the development of new techniques for applying and curing surface coatings. By far the most important of the 'environmentally compliant' technologies to emerge were cross-linked coatings and printing inks based on oligomeric bis-acrylate monomers. These could be cross-linked rapidly by UV light in the presence of UV sensitisers. Typical examples are the *polyester acrylates* (PEAs) and *poly(urethane) acrylates* (PUAs). The molecular weights of the oligomers have to be carefully controlled to give the correct viscosity for surface application.

$$CH_2{=}CHCO(CH_2)_6[OC(CH_2)_4CO(CH_2)_n]OCCH{=}CH_2$$

PEA

$$CH_2{=}CHCO(CH_2)_2O[CNH(CH_2)_6NHCO(CH_2)_4O]_nCCH{=}CH_2$$

PUA

For some industrial applications water-borne coatings and 'powder' coatings have been developed. The latter are applied at high temperatures to metal surfaces and dispense completely with low M_r precursors and solvents if higher temperatures can be tolerated.

Polymers in Building and Civil Engineering

A very visible and valuable contribution of polymers in the building industry is the replacement of wood in window frames and in outdoor cladding. The advantage of plastics materials is their resistance to biodegradation and this characteristic, coupled with reduced decoration costs, makes them the materials of choice as replacements for wood and iron. Pigmented rigid (unplasticised) PVC (UPVC) is the most widely used polymeric material for outdoor use but it has to be very effectively stabilised against the effects of the weather by the use of synergistic light stabilisers (Chapter 3). The main effect of the weather on white (TiO_2) pigmented UPVC is to reduce surface gloss, which is due to the gradual erosion of the surface layers by photooxidation. Fortunately, 'chalking' as surface erosion is called, actually protects the polymer bulk from rapid deterioration by screening it from the effects of UV light.

PVC is also used in windows and rooflights. Transparent PVC is much less durable than the white pigmented form described above. As a result of photooxidation, catalysed by UV light, the impact resistance steadily decreases over a period of several years with associated discolouration. UV absorbers have some protective effect but complex synergistic combinations of antioxidants and stabilisers are used to achieve a cost effective lifetime (Chapter 3, page 57). Polycarbonates are also used in structures where good clarity and toughness are required.

Polymers are used in much less obvious but often very important engineering applications. A major 'hidden' use of polymers is as adhesives. The traditional 'glues' were of course polymers derived from natural products. Some of the best of these still survive but synthetic polymer chemistry had led to very significant advances, particularly in the development of 'hot-melt' adhesives that do not require solvents for their application. Another important development is the 'reactive' adhesives based on ethyl cyanoacrylate:

$$CH_2{=}C\underset{CO_2CH_2CH_3}{\overset{C{\equiv}N}{<}}$$

Ethyl cyanoacrylate

which are relatively stable in the presence of air but which cross-link rapidly in the absence of air and in the presence of a radical generator to give 'superglue'. This type of adhesive is now widely used in engineering as 'thread-lockers' for the traditional nut and bolt but such is the adhesive qualities of the cyanoacrylates and related compounds that superglues may eventually replace nuts and bolts altogether in the joining of materials.

Rubber retains its resilience much better under compression than under extension and for this reason, when appropriately stabilised with antioxidants, it can be used in bridge bearings and load-bearing foundations in buildings. Another not immediately obvious use of polymers is in cavity insulation. A major advantage of polymers is that they can be formed as solid foams inside cavity walls of existing buildings without having to dismantle the wall. This can be done by injecting a liquid mixture of the component precursors of a foam together with a catalyst for the chemical reaction. By adjusting the concentration of the catalyst the kinetics of polymerisation is made to match the expansion of the foam into its final cross-linked state. Although this is a very cheap and convenient way of insulating existing houses, it has to be done very carefully to make sure that the low M_r and sometimes toxic chemicals used in the polymerisation are completely reacted and that degradation products cannot subsequently migrate into the human environment.

Polydiene rubbers were used for many years in low voltage electrical insulation because of their low electrical conductivity and flexibility. However, the traditional rubbers degrade by peroxidation relatively rapidly even when stabilised with antioxidants. This frequently leads to fires due to electrical shortage and, in recent decades, they have been largely replaced with saturated elastomeric materials (*e.g.* EP rubbers and plasticised PVC) which are much more durable.

Plastics in the Public Utilities

Polymers have in recent years assumed an increasingly important role in underground applications. These include piping, ducting and underground chambers where previously steel or concrete would have been used. New uses include impermeable membranes in the containment of water in reservoirs and of effluents in sanitary landfill (Figure 2.3), in grids and nets in soil stabilisation and in underground electricity cables. These sub-soil uses of polymeric materials makes use of their resistance to biodegradation and they have been given the general description 'geopolymers' or 'geosynthetics'.

The underground transport of oil, water and gas by pipeline is an ever increasing aspect of utility supply to industrial and domestic destinations. Iron and steel were the main materials of pipe construction 50 years ago and as they fail due to corrosion they are now replaced by plastics that do not corrode. Particularly favoured are high density polyethylene (HDPE), linear low density polyethylene (LLDPE), polypropylene (PP) and to a lesser extent rigid PVC. These materials give very adequate 'hoop' strength at the pressures used in the transport of liquids and natural gas that are conveyed in relatively large diameter pipes. It is relatively easy to test the bursting pressures of pipes when they are initially manufactured but it is extremely difficulty to predict whether geopolymers will survive without failure for the 50 years design lifetime that is normally required. There is a fundamental problem with geosynthetics in that none of these materials have been used for this period of time under actual service conditions and still less have they been tested to destruction under conditions of long-term use. It is necessary then to rely on the extrapolation of burst test data from *accelerated tests.* This essentially means increasing the temperature of the test and hoping that the data can be linearly extrapolated to ambient temperatures. Figure 2.4 shows typical test data for 'hoop-stress' decrease for PP pipes at different temperatures.

It is clear that there is a sharp time-dependent change in ageing behaviour that changes with test temperature and similar results have been found with LLDPE. It is not possible to predict with certainty from accelerated tests how long it will take for this change to occur at 20 °C and what will be the subsequent rate of change. It seems possible that discontinuity in the curves corresponds to the depletion of the antioxidants in the pipe, which may be due to physical as well as

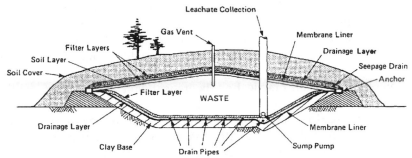

Figure 2.3 *The use of polymers in membrane liners in sanitary landfill*
(Reproduced with permission from H. E. Haxo and P. D. Haxo, in
Durability and Aging of Geosynthetics, ed. R. M. Koerner, Elsevier
Applied Science, 1989, p. 34)

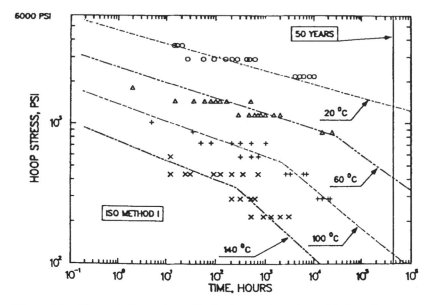

Figure 2.4 *Change in 'hoop strength' of polypropylene pipes with time of
accelerated ageing*
(Reproduced with permission from A. E. Lord and Y. H. Halse, in
Durability and Aging of Geosynthetics, ed. R. M. Koerner,
Elsevier Applied Science, 1989, p. 309)

chemical processes. The physical loss of antioxidants depends upon the temperature and the environment, which includes the nature of the contacting medium. Both aqueous and organic media are solvents for antioxidants and in this situation, antioxidant diffusion to the surface of the pipe becomes a controlling factor in its durability. Since there is at present insufficient practical or theoretical data to make firm predictions about the durability of geosynthetics, a worst-case scenario has to be assumed. However, theoretical and practical considerations suggest that conventional low M_r additives should be replaced by more *substantive* systems that will not be physically lost from the polymer long before the end of the design life of the component. Techniques for increasing the substantivity of antioxidants and stabilisers in polymers will be discussed in Chapter 3.

A major advantage of plastic piping is that it can be inserted into an existing underground channel, which may be an earlier metal duct that has fractured through corrosion. However, great care has to be taken that the new pipe is not distorted by obtrusions since these introduce stresses into polymers which induce accelerated fracture. This phenomenon, known as 'environmental stress cracking' (ESC), is accentuated by solvents. ESC is not primarily a peroxidation phenomenon since it does not occur uniformly through the polymer bulk. It is caused by the development of surface imperfections which accelerate the propagation of cracks into the undegraded polymer. However, stress-cracking can be initiated by surface photooxidation and care has to be taken to avoid the exposure of pipes to sunlight before burial in the ground.

Plastics are also now used as a replacement for concrete in underground chambers. The performance of chambers is much less critical than that of pipes since they are not under pressure and can be readily replaced when necessary. Increasingly, recycled polyolefins are being used in this kind of application, which is a particularly appropriate 'second use' for mixed thermoplastic polymers that are not exposed to UV light (Chapter 4, page 86).

The above phenomena are also relevant to the long-term behaviour of polymers used in underground membranes (Figure 2.3). In the case of sanitary landfill it may take many decades for biodegradation to be complete and during this time leachates will be produced continuously. It is therefore essential that these do not escape into water courses through cracks or tears in the bottom membrane. The integrity of underground containers over a long period becomes even more important in the case of low-level radioactive waste, which is frequently stored in below-ground vaults. In this application, it is equally important that water does not permeate the structure from outside. Cross-linked

HDPE is the preferred material of construction for underground 'high integrity' containers but these have been used for only about 20 years and there must be considerable doubt about their ability to resist the effects of high energy radiation in the presence of air over the next 20 years.

Geotextiles, particularly woven polypropylene, are used in soil stabilising networks beneath road surfaces and durability of road surfaces in recent years is largely due to these polymer-based foundations. Although not exposed to UV light they generally contain up to 2.5% carbon black to reduce sensitivity to applied stresses. Very fine particle size carbon (16–20 nm) have to be used for optimal durability and mechanical performance. Another environmental application of polyolefin geotextiles is in landfill leachate filtration systems which have to survive for very many years. However, the major problem here is the microbial blocking of the textile filters rather than their environmental degradation.

In the environs of roads, highways and in carparks, geogrids and geotextiles are again used to stabilise the soil (for example in sloping embankments and on flat surfaces subjected to heavy wear). Since these are at least partly in sunlight, the polymers are much more susceptible to environmental degradation, light, oxygen, mechanical stress and environmental pollutants, all of which accelerate the peroxidation processes that lead to degradation. There is an increasing tendency, owing to the need to find uses for waste plastics, to use recycled materials in these applications and unless the structures are temporary, as in the case of traffic cones *etc.*, then serious problems in maintenance can be anticipated due to premature mechanical failure.

Permanent soil stabilisation structures in underground membranes for landfill sites or reservoirs may require lifetimes of 75 to 100 years and durability becomes a critical aspect of the material design. An empirical 'trial and error' approach is still practiced by geosynthetics technologists. This approach was justified in the past and has provided some valuable experimental data, but it is clearly not the preferred present-day option in major construction projects where long-term durability is critically important. The same criteria apply here as in the case of pipes. Since it is not yet possible to predict the long-term durability of the polymer from accelerated tests, the best available scientific understanding of polymer degradation and stabilisation must be employed, including the 'anchoring' of antioxidants in the polymer matrix.

Polyolefins are now being used increasingly in underground electricity transmission. Although not normally called geosynthetics in this application, the same durability characteristics are essential here. One of

Figure 2.5   *XLPE insulated underground cable*
(Reproduced with permission from *BICC Supertension Cables*,
publication 933, 1981)

the success stories in electrical cable manufacture has been cross-linked
polyethylene (XPLE) cable for high capacity underground electricity
transmission (Figure 2.5).

Cross-linking of the outer casing is achieved by grafting a vinyl silane
to polyethylene by means of a peroxide as part of the cable extrusion
process. The chemistry is shown in Scheme 2.1. When the cable is
subsequently soaked in water, the alkoxy silanes hydrolyse and cross-
link within the matrix. It is an unresolved question whether polymers
that have been modified by a free radical mechanism will survive for the
expected lifetime of underground cables (*ca.* 75 years).

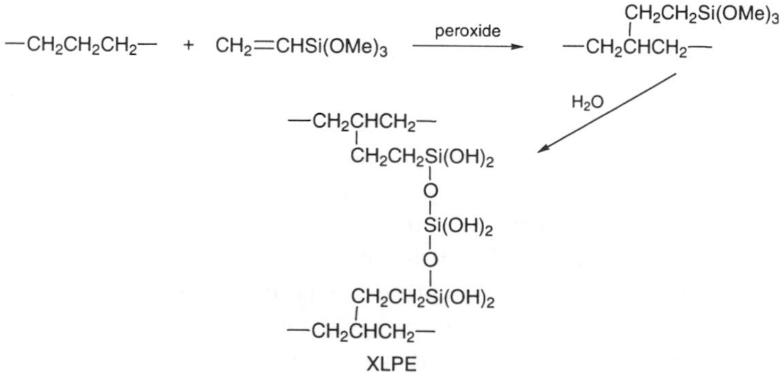

$$-CH_2CH_2CH_2- \quad + \quad CH_2{=}CHSi(OMe)_3 \xrightarrow{\text{peroxide}} \begin{array}{c} CH_2CH_2Si(OMe)_3 \\ | \\ -CH_2CHCH_2- \end{array}$$

$$\Big\downarrow H_2O$$

$$\begin{array}{c} -CH_2CHCH_2- \\ | \\ CH_2CH_2Si(OH)_2 \\ | \\ O \\ | \\ Si(OH)_2 \\ | \\ O \\ | \\ CH_2CH_2Si(OH)_2 \\ | \\ -CH_2CHCH_2- \\ \text{XLPE} \end{array}$$

Scheme 2.1 *Silane cross-linking of polyethylene*

The initial mechanical properties of XLPE are exceptionally suitable for
electrical cable. Because it is cross-linked, it will withstand relatively
high temperatures without undergoing 'creep' while at the same time it
remains tough and resistant to fracture at relatively low temperatures.

However, considerably more experience is required on the long-term durability of XLPE before this can be considered to be a completely successful process. It is likely that new stabilisation systems will be required that are not adversely affected by the free radical generators used in the fabrication process (Chapter 3).

Overhead electrical conductors with their associated pylons are now viewed with some disfavour by planning authorities because of their visual impact on the environment and health hazards associated with their magnetic fields. Consequently there is considerable activity, particularly in the USA, in developing efficient underground transmission systems. The major problem with all electricity transmission is the linear loss of electrical energy with distance, energy which is wasted as heat in the environment. A long-term objective is to reduce energy loss to a minimum by producing and using electricity and associated heat locally in small power stations (combined heat and power, CHP). In the meantime, the electricity utilities are investigating new materials with improved dielectric properties to replace the traditional oil-filled paper laminates, which had a relatively high dielectric loss, by polymer–paper laminates. Polypropylene paper oil-filled laminates (PPL OF) have been found to be particularly successful since they reduce dielectric losses by 70% compared with oil-filled paper. This amounts to a very considerable saving in energy in the transmission of high powered (*e.g.* 400 kV and above) electricity over long distances. Furthermore, the compact cable configuration achieved by the use of PPL OF minimises electromagnetic fields which are believed to be a cause of leukaemia in populations living in the vicinity of overhead transmission lines.

Polymers in Biology and Medicine

The biological inertness and lightness of polymers make them very attractive in potential biomedical applications. Typical examples of this are in dental applications where UV cross-linking is playing an increasing role. 'Soft' hydrogels that absorb water and transmit oxygen are replacing hard polymers in contact lenses. Replacement plastic prostheses are now state of the art. However, in this application the durability and biocompatibility of the polymer under the aggressive conditions to which they are subjected in use is a basic design parameter. It is known that low M_r antioxidants in polymers are slowly leached by body fluids. This is undesirable for two reasons: firstly they may be toxic to the host organs and, secondly, because the removal of antioxidants will cause premature biodegradation and mechanical failure. The technology for attaching antioxidants and stabilisers to polymers has already been developed but has not so far been translated into the specialised area of biomaterials.

Another use for polymers in the body is in temporary inclusions such as sutures and supporting meshes which are required to dissolve and biodegrade over a relatively short period of time. This requires the design of new material which, unlike the present range of bio-inert polymers, can then be bioassimilated into the body after they have served their purpose. This will be discussed in Chapter 5.

THE PUBLIC IMAGE OF 'PLASTICS'

Plastics packaging has over the years had a very bad press. In spite of the contribution of plastics to food hygiene, energy conservation and general amenity, they have come to be regarded as 'squandered' mineral resources. They are perceived to be non-biodegradable and persistent in the environment when discarded. In permanent applications plastics are considered to be inferior substitutes for 'quality' materials like wood, metals and ceramics. Furthermore, in furniture and upholstery, they are known to be highly flammable and the cause of fires in domestic and commercial buildings.

The negative aspects of plastics are rarely discussed in the context of their advantages in energy conservation by the 'green' movement. It is a false assumption that 'natural' materials from renewable resources must inevitably be more ecologically acceptable than synthetics but this generally underlies the rejection of plastics and some would replace plastics in packaging by traditional materials without considering the ecological implications. It is not surprising then that the concerned public gains a one-sided view of the environmental impact of plastics.

Polymer industry PR has so far failed to provide a balanced alternative view to that put forward by the 'green' lobby. In the 1970s it was suggested by reputable scientists like James Guillet of Toronto University that the plastics litter problem could be effectively mitigated by making the selected plastics packaging materials degradable. Organisations such as the Society for Plastics Industry (SPI) in the USA and British Plastics Federation (BPF) and the Industry Council for Packaging and the Environment (INCPEN) in the UK rejected without adequate consideration and trial the concept of induced degradability as a contribution to litter control (Chapter 5).

Waste and litter is not the only concern of the public about plastics. There is enough truth in public criticisms of the performance of many household plastics products to concern the polymer industries. It is perhaps significant that rubber-based products such as motor car tyres, which have evolved over a much longer time, are not seen in the same way by the public. It is important to ask whether the public image of plastics is based on prejudice or whether some of the criticisms are real and within the capability of the industry to remedy. Many common household articles made from plastics give the impression that they have been designed for cheapness rather than quality. Consumer products made from plastics are frequently found to deteriorate much more rapidly in use than traditional products. The garden watering can or patio furniture may last for a very few years compared with metal or wooden equivalents manufactured scores of years ago and still in use.

The user is left with the impression that plastics components have not been thoroughly tested in the working environment before being put on the market. Very often very complex equipment fails due to the deterioration of a relatively small but nevertheless vital polymer-based component.

Durability is given a very low priority at the component design stage. It is evident that many materials technologists do not fully appreciate the environmental limitations of modern polymers and in particular the importance of correct stabilisation. This severely limits their ability to choose between competing materials and the problem is compounded by the reticence of many polymer additive manufacturers to disclose to the converters the reason for their own recommended formulations. The principles underlying the design of polymer-based product for long-term performance will be discussed in the next chapter.

FURTHER READING

1 K. D. Weiss, Paint and Coatings: A mature industry in transformation, *Prog. Polym. Sci.*, 1997, **22**, 203–245.
2 *Durability and Aging of Geosynthetics*, ed. R. M. Koerner, Elsevier Applied Science, 1989.
3 *Polymers and Ecological Problems*, ed. J. E. Guillet, Plenum Press, 1973.
4 D. Birkett, Reactive adhesives, proactive chemistry, *Chem. Br.*, 1998, **34**, 20–24.
5 J. Guillet, Plastics and the environment, *Degradable Polymers: Principles and Applications*, ed. G. Scott and D. Gilead, Chapman & Hall, 1995, Ch. 12.

Chapter 3

Environmental Stability of Polymers

POLYMER DURABILITY: A DESIGN PARAMETER

It is clear from Chapters 1 and 2 that the introduction of polymeric materials into consumer products has resulted in reduced energy utilisation during manufacture and use. It is not quite so obvious that extending the useful lifetime of domestic and industrial equipment is another and equally important means of energy conservation. However, this concept conflicts with the ethos of the 'throw-away' society where fashion rather than ecology normally governs the lifetime of consumer products. Substantial profits are made through premature obsolescence and the sale of replacement parts is very often more profitable than the sale of the original equipment. Consequently, the manufacturing industry does not have a strong incentive to improve the durability of its products; such improvement adds a small amount to initial cost but in the long-term it must be beneficial to the environment. It is ultimately the responsibility of society to decide whether energy and materials conservation are sufficiently important to require industry to change the current market philosophy.

From a more practical standpoint, doubling the life of domestic equipment could save the average householder many hundreds of pounds a year and would at the same time reduce national energy consumption by millions of kilowatt hours. Polymer components normally fail either through bad engineering design, bad materials design or a combination of both and it is rarely possible for the non-technical user to see which of these is responsible. For example, a hook on a vacuum cleaner may break either because it was not initially designed to withstand the stresses and strains in everyday use or because the properties of the material have changed during use. Most frequently the cause is the latter not the former.

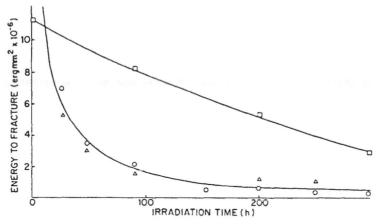

Figure 3.1   *Effect of photooxidation on falling weight impact resistance of PVC*
(□) *and ABS* (10%) *modified PVC* (○, △) *duplicate results*
(Reproduced with permission from G. Scott, in *Development in*
Polymer Stabilisation – 1, ed. G. Scott, Applied Science
Publications, 1979, p. 310)

The composition of polymeric materials is an important determinant
of durability and some serious mistakes were made during the last two
decades through a lack of understanding of the principles underlying
polymer deterioration. A typical example was the development of the
'impact modified' polymers in which oxidatively unstable rubbers (*e.g.*
polybutadiene) were blended with or grafted to relatively light-stable
polymers such as polystyrene–acrylonitrile. The resulting acrylonitrile–butadiene–styrene (ABS) terpolymers (Chapter 1, page 12) showed
a marked initial improvement in toughness compared with the rubber
unmodified material but the rubber component had a disastrous effect
on durability. Rubber-modified plastics were also used as compounding
ingredients in otherwise durable polymers such as rigid PVC with
similar effects. Figure 3.1 shows the effect of adding ABS to PVC as a
toughening agent. The initial impact resistance is much superior to that
of unmodified PVC but, on exposure to light, the position rapidly
reverses due to photooxidation of the rubber phase in the impact
modifier.

Recently, more light-stable rubber modifiers have been developed
based on saturated rubbers but the rubber phase still remains the
'Achilles heel' of impact modified plastics and great care is required in
formulating polymers for environmental stability if they are to be used
in long-life products.

At the other end of the durability scale there is the problem of overstabilisation. Until recently, very little thought was given to the design of polymeric materials that end up in the waste stream or as litter but now, owing to the intervention of governments, 'environmentally friendly' materials are demanded. However, to return to traditional materials is to ignore the very considerable benefits in hygiene and energy utilisation that have resulted from the replacement of paper by plastics in food wrapping or in the transport of chemicals in plastics rather than in metals or glass. The polymer scientist is thus left with the task of designing polymeric materials for much more specific end uses than in the past.

ASSESSMENT OF POLYMER DURABILITY

It is relatively easy to mechanically test components to failure in the laboratory. Testing specifications and standards are laid down for the mechanical behaviours of components immediately after manufacture. However, as discussed in the previous chapter, it is unsafe to assume that these will remain unchanged over the lifetime of the component since the durability of a polymeric materials depends on its structure, on the environment in which it is used and on the stabilisation system used.

Many thousands of computer hours have been spent attempting to devise 'models' for polymer degradation that will allow the lifetime of a component to be predicted. This approach, in principle, requires a detailed knowledge of the rate constants of every chemical reaction involved in peroxidation and inhibition as well as the rates of physical loss of additives from polymers. However, rate constants for reactions known to occur in polymers may have no analogy to those occurring in classical chemistry. For example, mechanochemical degradation occurs during both the manufacture and use of polymer components and techniques are not yet available for quantifying these. Furthermore, a complete analysis of the effects of stress initiation is required before a complete mathematical description can be attempted. It is perhaps not surprising then that even the most sophisticated computer models do not yet permit the prediction of the durability of engineering components.

Testing to failure under service conditions is time consuming and costly and consequently most component manufacturers rely on information from polymer suppliers. This at best can only be a guide to component performance and the polymer technologist who wishes to know more about the environmental behaviour of his material must rely on lifetime prediction from accelerated tests. These are of two types, the first is a 'heat-ageing' test in which the change in mechanical properties of the polymer is followed at an elevated temperature to accelerate the process. The second is a light stability test in which the polymer is exposed to selected wavelengths of incident UV light (generally a xenon arc) and changes in mechanical properties (tensile strength, elongation at break and impact resistance) are measured.

In principle, measurement of failure time at two or more elevated temperatures should permit extrapolation to service temperature by means of the Arrhenius equation. In practice extrapolation from relatively modest accelerated temperatures to ambient temperature is very unsafe because it cannot be assumed that the Arrhenius relationship is linear. As was seen in the section on Plastics in the Public Utilities (page 30), discontinuities may occur in relevant property–time relationships which invalidate this procedure (Figure 2.4). Even less reliable is the commonly used comparison of oxidation induction times by differential scanning calorimetry for different stabilisers, these very rarely correlate with behaviour at lower temperatures and this type of test is best limited to informing the technologist whether an antioxidant is present. High temperature tests in forced air or oxygen does provide information about the volatility of the antioxidants under these defined conditions but it is rarely possible to extrapolate this to the environmental conditions in the vicinity of the polymer during service.

For many domestic and industrial uses of polymers, heat ageing is less important than exposure to light and environmental pollutants; however, where high temperatures are experienced in service, heat ageing tests are normally included in International Standards Organisation (ISO) tests. A major cause of failure of domestic and office equipment manufactured in polymeric materials is light-catalysed oxidation and although this is normally called 'weathering' it does not necessarily involve exposure to the outdoor environment. There are a number of outdoor exposure sites in various climates throughout the world where materials can be sent to determine the effects of the weather on polymer durability. Needless to say, these are generally in parts of the world where UV incidence and ambient temperatures are high (*e.g.* Florida, Arizona, Australia). There are other exposure sites in the Far East, which range from humid semi-tropical to dry and cool. Measurement of changes in mechanical properties at periodic intervals of samples exposed to different environments provide a useful guide to the weathering behaviour of polymers in most environments. There are some applications of plastics for which testing out-of-doors under conditions of use is essential. This applies particularly to agricultural films and 'disposable' packaging which have to last for a specified time (Chapter 5, page 110).

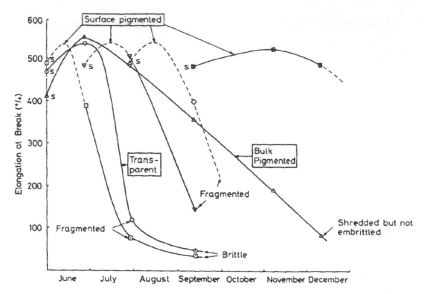

Figure 3.2 *Mechanical property change of photo-biodegradable polyethylene*
carrier bags during outdoor exposure in England
(Reproduced with permission from D. Gilead and G. Scott, in
Developments in Polymer Stabilisation – 5, ed. G. Scott, Applied
Science Publications, 1982, p. 103)

Outdoor lifetimes are days or months rather than years for more durable products (Figure 3.2).

Outdoor tests are expensive as a means of routine monitoring of
polymer durability. Furthermore they are subjected to the vagaries of
incident light and temperature which vary with the season (Figure 3.3).
Such variations may be misleading unless they are corrected by parallel
monitoring of incident light and other environmental components such
as humidity, temperature and atmospheric pollution, which play a
secondary but often synergistic role together with UV light in the rate of
environmental degradation of polymers.

Accelerated UV tests provide an alternative and generally much more
rapid means of measuring the effects of light on polymers under consist-
ent and reproducible conditions. The UV light is generally provided by
xenon arc lamps and some modern UV exposure cabinet provide a very
rapid assessment of polymeric materials (Figure 3.4) and correlate well
with environmental exposure. Any element of the environment can be
incorporated into a 'weatherometer'. For example, relative humidity
may be varied and an alternating water spray can be incorporated to
simulate rain. As has already been noted this surface treatment may

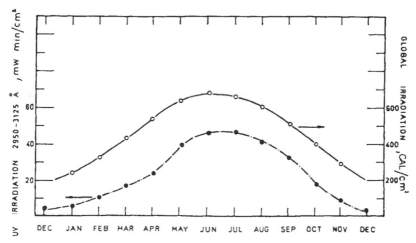

Figure 3.3   *Average daily sunlight distribution throughout the year (Israel)*
(Reproduced with permission from D. Gilead and G. Scott, in
Developments in Polymer Stabilisation – 5, ed. G. Scott, Applied
Science Publications, 1982, p. 100)

Figure 3.4   *Apparatus for accelerated photo-ageing of polymers (SEPAP
12/24).*
(Photograph courtesy of Professor J. Lemaire, University of
Blaise Pascal)

have a marked influence on the loss of antioxidants and stabilisers, particularly in thin films and fibres where the surface area to volume ratio is high.

Accelerated UV tests are normally conducted slightly above ambient temperatures (30–35 °C) and are generally more useful for assessing the durability of domestic and office equipment than outdoor tests. However, the technique used has to be calibrated with long-term performance in the user environment.

In the case of rubbers, other environmental agencies are frequently as important as light, if not more important. In addition to heat-ageing tests of the type described above, static strain tests in the presence of very low concentrations of ozone which are often found in industrial atmospheres are carried out to simulate the effects of the environment on rubber under stress. This phenomenon was first recognised as a serious problem in the 1939–1945 war when army vehicles were allowed to stand in 'combat readiness' for weeks or even months. Frequently the tyre side-wall in proximity to the road was found to be cracked, sometimes through to the tyre cord, whereas the unstrained segments of the tyre showed no obvious sign of degradation. Ozone rapidly cuts through strained rubber and it is a tribute to the rubber technologists' art that chemicals have been discovered which are able to compete with rubber for ozone while at the same time forming a protective skin in the rubber surface. A second type of degradation due to fatigue and often referred to as *flex-cracking* occurs in rubbers subjected to dynamic (oscillating) stress. This is often evident in the grooves between tyre treads and is primarily due to stress initiated oxidation (mechanooxidation) which at the molecular level is the same process that leads to mechanooxidation of polymers during processing. Ozone exacerbates this process but, unlike ozone cracking, it is not the primary cause.

CHEMICAL ASPECTS OF POLYMER DEGRADATION

Polymer *degradation* is the change that occurs in the chemical composition of a polymer with time due to the effects of heat, light or mechanical stress. Degradation is almost always accompanied by corresponding mechanical or aesthetic changes which may make polymer artifacts unsuitable for their original purpose. *Durability* is the converse of degradability. It is measured practically as the time required for the irreversible loss of the useful properties of the polymer and this in turn is closely dependent on chemical changes in the polymer. As already noted, extending the lifetime of consumer goods is an important practical way to reduce both energy and materials utilisation of consumer products. In practice, and to the detriment of the environment, fashion rather than ecological considerations frequently determines the lifetime of clothing, cars and household commodities. The fundamental conflict between modern fashion driven consumerism, with associated industrial growth, and materials and energy conservation cannot be solved by scientists and technologists. It is ultimately a social problem that only governments can address but it is the responsibility of the scientist to indicate to politicians how materials design can play a part in conservation.

Polymers can undergo degradation in different ways depending on the environment. In the absence of air or water *anaerobic* degradation or pyrolysis occurs. By contrast *aerobic* degradation involves *peroxidation* which generally occurs at much lower temperatures than pyrolysis and since peroxidation is catalysed by light and metal contaminents, it is the most important process involved in the environmental degradation of carbon-chain polymers. Commercial hetero-chain polymers such as the polyesters, polyamides and polyurethanes are also relatively stable to pyrolysis but they degrade in the environment by a combination of peroxidation and hydrolysis.

Thermal (Anaerobic) Degradation

Although thermal degradation of the common polymers generally occurs well above the temperatures encountered during normal service, there are abnormal situations such as fire where the thermal degradation temperature of most polymers is exceeded and where volatile organic compounds are formed by *pyrolysis* (or *thermolysis*). The flammability of organic polymers in the fire environment is a major hazard which has become increasingly significant in recent years due to the replacement of naturally occurring fibres such as wool, animal hairs, feathers, *etc.* in furniture by the more flammable polyurethane foams and synthetic textiles. The initial cause of fire is normally some external source like a cigarette or match which results in the formation of volatile fuels by thermal degradation of the high M_r polymer matrix. This in turn gives rise to a self-sustaining and rapidly propagating flame depicted in the combustion model shown in Scheme 3.1. The polymer itself is not a fuel until it has been pyrolysed to give volatile, flammable gases, so that if the heat is removed, for example by spraying with water, then fuel production ceases. Similarly, if the oxygen is removed from the fuel by blanketing the polymer with an inert gas or solid, then the oxygen and fuel will be diluted and the polymer will cease to burn. These are the normal ways in which established fires are dealt with by the fire-services. It is of fundamental importance, however, to inhibit or retard the initiation of combustion so that fire-brigade action is not necessary. A great deal of progress has been made in recent years to produce 'self-extinguishing' polymers. This has been accompanied by appropriate legislation to ensure that furniture fabrics and foams conform to what can now be practically achieved in commercial 'fire-retardant' materials. The design of flame-retarded polymers will be discussed below.

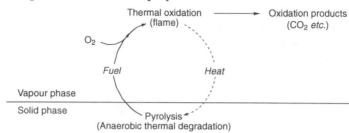

Scheme 3.1 *Polymer combustion model*

In recent years, the pyrolysis of polymers under carefully controlled conditions has become an important part of the strategy for recovering useful products from waste plastics. Thus polystyrene (PS) and poly-(methyl methacrylate), PMMA, begin to 'depolymerise' to their respective monomers above 300 °C. In the case of PMMA this occurs in almost quantitative yield. Unfortunately, very few polymers break down so cleanly to monomer by pyrolysis. PS gives only 42% of styrene and polyethylene (PE) and polypropylene (PP) give only 1 and 2% respectively. PVC gives no monomer at all but eliminates all its hydrogen chloride to give a black carbonaceous residue. Nevertheless pyrolysis of waste, which can itself be fuelled by burning part of the waste, gives rise to a range of low molecular weight hydrocarbons which are useful as fuel. This will be discussed further in Chapter 4.

Oxidative Degradation

Oxygen is omnipresent in our environment and the deterioration of rubbers and plastics by *peroxidation* is the normal cause of property deterioration in most polymers under ambient conditions. Peroxidation is a free radical chain reaction, shown in summary in reactions 3.1 and 3.2. Under normal conditions it is initiated by *hydroperoxides* that are formed in each cycle of the peroxidation chain sequence (reaction 3.1). Hydroperoxides are very unstable compounds due to the weakness of the peroxide bond which readily undergoes *thermolysis* when heated (reaction 3.3). This reaction is powerfully catalysed by transition metal ions,

Peroxidation radical-chain reaction

$$PH + POO\cdot \longrightarrow P\cdot + POOH \qquad (3.1)$$

$$P\cdot + O_2 \longrightarrow POO\cdot \qquad (3.2)$$

Initiation reactions in peroxidation

$$2POOH \xrightarrow[\text{(M}^+/\text{M}^{2+})]{\text{heat}} PO\cdot + POO\cdot + H_2O \qquad (3.3)$$

$$ROOH \xrightarrow[\text{(M}^+/\text{M}^{2+})]{hv} RO\cdot + \cdot OH \qquad (3.4)$$

$$P\text{--}P \xrightarrow{\text{shear}} 2P\cdot \xrightarrow{O_2} 2POO\cdot \xrightarrow{PH} 2POOH \qquad (3.5)$$

PH, P–P represent polymers

which are the main initiators of peroxidation in the absence of light. *Photolysis* of hydroperoxides is the main initiation process in the outdoor environment (reaction 3.4) and sometimes it is important even in artificial light. Peroxidation of commercial polymers begins almost as soon as the polymer comes from the manufacturing process but it is accelerated by the processing operation in which the polymer is subjected to high shearing forces in the viscous liquid state. Owing to the presence of a small amount of oxygen which cannot be easily eliminated under technological conditions, this process, which is known as *mechanooxidation*, leads to the formation of peroxyl radicals and hydroperoxides (reaction 3.5). The latter remain in the polymer unless removed by antioxidants (page 57) and residual hydroperoxides are the main cause of subsequent environmental degradation. This is particularly important in the case of films and fibres where the high surface area to volume ratio leads to rapid surface peroxidation in light. Mechanical shear is also a very important initiator in rubber goods, notably tyres, and this process causes *'fatigue'*. Durability is thus a critical factor in the design of materials used in engineering components.

POLYMER PROCESSING

Thermoplastic polymers, as they come from the chemical manufacturing process, are powders or pellets. To convert them into fabricated products for the market place, they have to be first heated to high temperatures in an internal mixer or *screw extruder* (Figure 3.5). This procedure permits the incorporation of additives that are essential to the subsequent service performance of the fabricated product. The screw also conveys the molten polymer to the die for the thermoforming process.

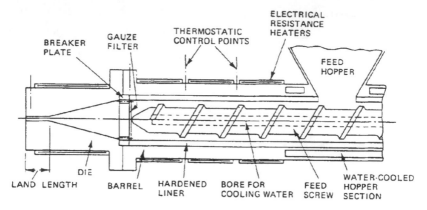

Figure 3.5   *Screw extruder*
(Reproduced with permission from J. A. Brydson, *Plastics*
Materials, Newnes-Butterworth, 1975, p. 144)

Antioxidants and stabilisers are added to plastics during extrusion to
provide rheological stability to the polymer during conversion and to
enhance durability during subsequent service. The commodity polymers, as noted above, are thermally stable enough not to undergo
pyrolysis to a significant extent during processing although in the case
of some speciality biodegradable polymers this can be a problem (Chapter 5) and represents a limitation on their industrial use. However,
processing thermoplastic polymers in a screw extruder is not, as might
appear at first sight, an anaerobic process. It is difficult and costly to
eliminate all oxygen from the polymer during conversion and some
peroxidation always occurs. A small amount of oxygen can do a great
deal of damage to polymer molecules during processing since high
temperature mixing of viscous polymers causes scission of highly
strained polymer molecules. These carbon-centred macroradicals are
highly reactive and react instantaneously with oxygen in the polymer.
High-shear mixing causes mechanooxidation which has been put to
positive use for over a 150 years in the manufacture of rubber products.

   Natural rubber, as it came from the tree, has a very high M_r and there is
no practical way of shaping it to tyres tubing, sheet, *etc.* or of incorporating additives (antioxidants, carbon black, sulfur and accelerators for
vulcanisation) without breaking down the macromolecule to smaller
segments. The early rubber technologists recognised that mechanical
'working' rubber led to plasticisation and they imaginatively described
this process as 'mastication'. A machine to do this, invented by Thomas
Hancock around 1830 and known as 'Hancock's pickle' (Figure 3.6) was

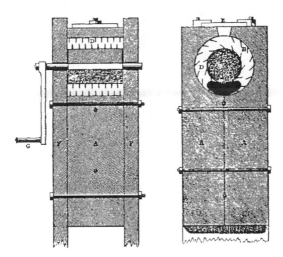

Figure 3.6 '*Hancock's Pickle*'. AA,F, *wooden body*; B, *stationary pins*;
D, *rotating drum with inserted pins*; E, *covered opening*; G, *handle*.
The rubber is sheared between the pins
(Adapted from H. J. Stern, in *Rubber Technology and Manufacture*,
ed. C. M. Blow, Butterworths, 1971, p. 2)

the precursor of the modern high-shear Banbury mixer (Figure 3.7) in
which high shearing forces are produced between the rotors powered by
electricity.

In this century it has been shown that 'mastication' of rubber pro-
duces free radicals with high chemical reactivity. Thus, for example,
macroradicals from *cis*-polyisoprene initiate vinyl polymerisation when
oxygen is rigorously excluded from a high shear mixer, giving rise to
block copolymers, typically containing rubber and methyl methacrylate
(Scheme 3.2). In the presence of oxygen this process is inhibited and the

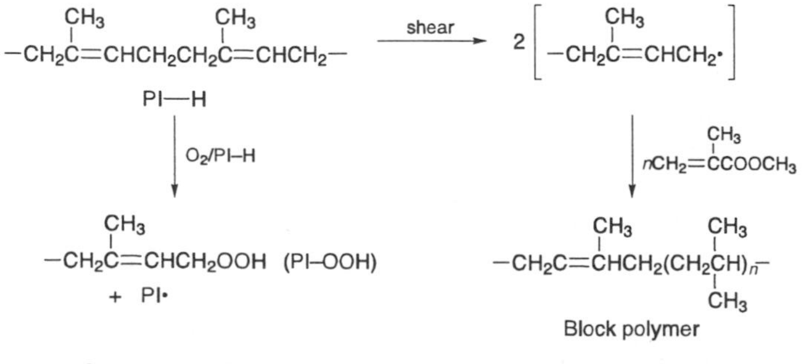

$$\underset{\text{PI—H}}{-CH_2\overset{\overset{\displaystyle CH_3}{|}}{C}=CHCH_2CH_2\overset{\overset{\displaystyle CH_3}{|}}{C}=CHCH_2-} \xrightarrow{\text{shear}} 2\left[-CH_2\overset{\overset{\displaystyle CH_3}{|}}{C}=CHCH_2\bullet\right]$$

$$\downarrow O_2/PI\text{–}H \qquad\qquad\qquad \downarrow nCH_2=\overset{\overset{\displaystyle CH_3}{|}}{C}COOCH_3$$

$$-CH_2\overset{\overset{\displaystyle CH_3}{|}}{C}=CHCH_2OOH \ (PI\text{–}OOH) \qquad -CH_2\overset{\overset{\displaystyle CH_3}{|}}{C}=CHCH_2(CH_2\overset{\overset{\displaystyle CH_3}{\underset{\displaystyle CH_3}{|}}}{C}H)_n-$$

$$+ \ PI\bullet$$

Block polymer

Scheme 3.2 *Mechanooxidation (mastication) of cis-poly(isoprene)*

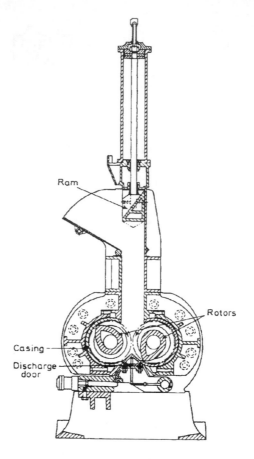

Figure 3.7   *Banbury mixer*
(Reproduced with permission from B. G. Crowther and H. M.
Edmondson, in *Rubber Technology and Manufacture*, ed. C. M.
Blow, Butterworths, 1971, p. 266)

molecular size of the polymer molecule is dramatically reduced with the formation of hydroperoxide groups at the end of the broken chain. This in turn results in the reduction of the viscosity of the rubber.

The process of raw rubber mastication involves a very high energy input and indeed this is by far the most important energy-consuming operation in the manufacture of the modern automotive tyre. To reduce energy usage and cost, chemists have developed additives to 'plasticise' rubber during mastication. Sulfur compounds such as pentachloro-thiophenol (PCTP) or thio-β-naphthol (TBN) were found to catalyse

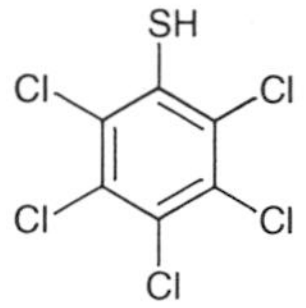

PCTP 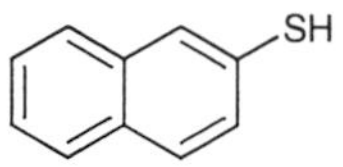   TBN

the formation of free radicals from the hydroperoxide (PI–OOH), thus producing more free radicals to initiate the peroxidation chain reaction. The induced decomposition of hydro-peroxides described in Scheme 3.3 leads to a substantial reduction in the usage of electrical energy due to the catalytic peroxidation, which results in molecular weight reduction. This process was called *chemical plasticisation* or '*peptisation*' by rubber technologists and, like the term 'mastication', 'peptisation' was borrowed from the analogous biological processes.

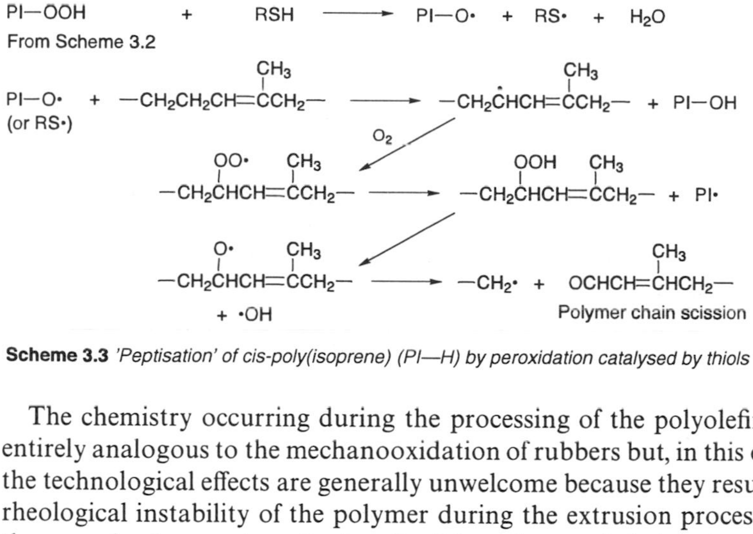

Scheme 3.3 *'Peptisation' of cis-poly(isoprene) (PI—H) by peroxidation catalysed by thiols (RSH)*

The chemistry occurring during the processing of the polyolefins is entirely analogous to the mechanooxidation of rubbers but, in this case, the technological effects are generally unwelcome because they result in rheological instability of the polymer during the extrusion process. In the case of polypropylene, the length of the polymer chain is reduced by the same chemical processes already described for *cis*-polyisoprene (Scheme 3.2). Polyethylene is quite different since the macroalkoxyl radicals formed by mechanooxidation react with macroalkyl radicals. The chemical result is molecular enlargement and the technological consequence is the formation of gel in the polymer (Scheme 3.4). This introduces physical weakness into fabricated artifacts, particularly in thin profiles such as films or fibres.

PVC behaves quite differently during processing. The initial step is the same, the formation of macroalkyl radicals, but the latter can undergo a rapid 'unzipping' of hydrogen chloride from the initial

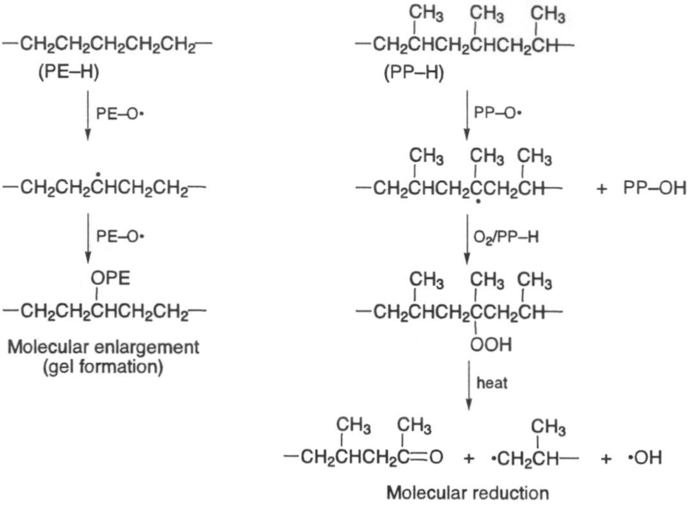

$-CH_2CH_2CH_2CH_2CH_2-$
(PE–H)

$\downarrow$ PE–O•

$-CH_2CH_2\overset{\bullet}{C}HCH_2CH_2-$

$\downarrow$ PE–O•

$\overset{\displaystyle OPE}{-CH_2CH_2CHCH_2CH_2-}$

Molecular enlargement
(gel formation)

$\overset{\displaystyle CH_3\ \ \ CH_3\ \ \ CH_3}{-CH_2CHCH_2CHCH_2CH-}$
(PP–H)

$\downarrow$ PP–O•

$-CH_2CHCH_2\overset{CH_3\ \ \ CH_3\ \ \ CH_3}{\underset{\bullet}{C}}CH_2CH-$ +   PP–OH

$\downarrow$ O$_2$/PP–H

$-CH_2CHCH_2\underset{\displaystyle OOH}{\overset{CH_3\ \ \ CH_3\ \ \ CH_3}{C}}CH_2CH-$

$\downarrow$ heat

$-CH_2CHCH_2\overset{CH_3\ \ \ CH_3}{C}=O$ +   •$\overset{CH_3}{CH_2CH}-$ +   •OH

Molecular reduction

Scheme 3.4 *Effects of processing on the structures of polyethylene (PE—H) and polypropylene (PP—H)*

radical, in competition with hydroperoxide formation (Scheme 3.5). Thus both hydrogen chloride and hydroperoxides are formed and they react to give free radicals that catalyse the further degradation of PVC. The technological consequence is the development of an intense discolouration of the polymer with relatively small change in molar mass, at least during the early stages. This is aesthetically unacceptable to polymer technologists but it is much more difficult to inhibit than the mechanical property changes that occur in other thermoplastic polymers and more complex stabiliser systems are required to prevent it.

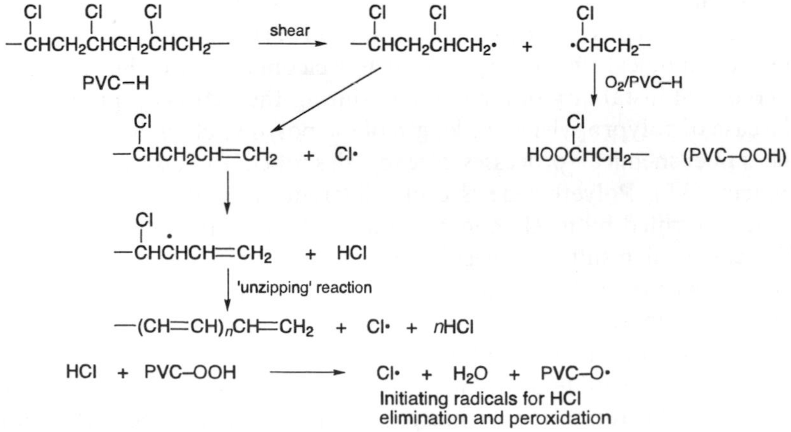

$-\overset{Cl}{C}HCH_2\overset{Cl}{C}HCH_2\overset{Cl}{C}HCH_2-$ $\xrightarrow{\text{shear}}$ $-\overset{Cl}{C}HCH_2\overset{Cl}{C}HCH_2•$ +   •$\overset{Cl}{C}HCH_2-$

PVC–H

$\downarrow$ O$_2$/PVC–H

$-\overset{Cl}{C}HCH_2CH=CH_2$ +   Cl•

HOO$\overset{Cl}{C}HCH_2-$ (PVC–OOH)

$\downarrow$

$-\overset{Cl}{\underset{\bullet}{C}}HCH CH=CH_2$ +   HCl

$\downarrow$ 'unzipping' reaction

$-(CH=CH)_nCH=CH_2$ +   Cl•   + nHCl

HCl   +   PVC–OOH $\longrightarrow$ Cl•   +   H$_2$O   +   PVC–O•
Initiating radicals for HCl
elimination and peroxidation

Scheme 3.5 *Mechanodegradation of PVC during processing*

Hydroperoxides formed during processing of polymers have a profoundly deleterious effect on the long-term performance of products made from them. This is a consequence of the very great sensitivity of hydroperoxides to light which leads to the rapid deterioration of the polyolefins, PVC and rubber-modified polymers in the outdoor environment. Their formation is of crucial importance to the 'weathering' of industrial polymers since these highly reactive free radicals produced are initiators for the *photooxidation* of polymers.

WEATHERING OF POLYMERS

All polymers degrade in the outdoor environment due to the initiating effects of impurities introduced during the manufacturing operations. The rate at which they do so varies widely, depending on their chemical structure. The presence of 'stabilisers' added during polymer conversion prevents rheological changes in the polymer both in the extruder and subsequently. For example, many thermoplastic polymers (*e.g.* polypropylene or PVC) without antioxidants or stabilisers become brittle within a few weeks on exposure to the outdoor environment, whereas a properly formulated pigmented polymer (*e.g.* UPVC, widely used in windows or as cladding on buildings) may last for decades. However, in all cases, the chemistry of their deterioration is very similar and involves the photolytic breakdown of hydroperoxides (reaction 3.4) which are always the first products to be formed (Scheme 3.6). Ketones also break down through the triplet excited states and both the Norrish type I and Norrish type II photolyses results in chain scission. However, only the Norrish I reaction gives free radicals and the ultimate products formed from these are carboxylic acids of much shorter chain length than the original polymer.

BIODEGRADATION OF POLYMERS

Low molecular weight dicarboxylic acids, keto acids and hydroxy acids have been shown to form as photooxidation products of polyethylene and polypropylene. These are almost certainly formed by intramolecular reactions of alkylperoxyl and peracyl radicals shown typically in Scheme 3.7. 'Back-biting' along the aliphatic chain gives rise to unstable hydroperoxides and the elimination of small molecular fragments. It will be seen in Chapter 5 that these low molar mass oxidation products, which are already present in the environment from natural sources, are the first point of microbial attack in the surface of environmentally degraded polymers, leading to oxidation initiated *bioerosion* (Chapter 5).

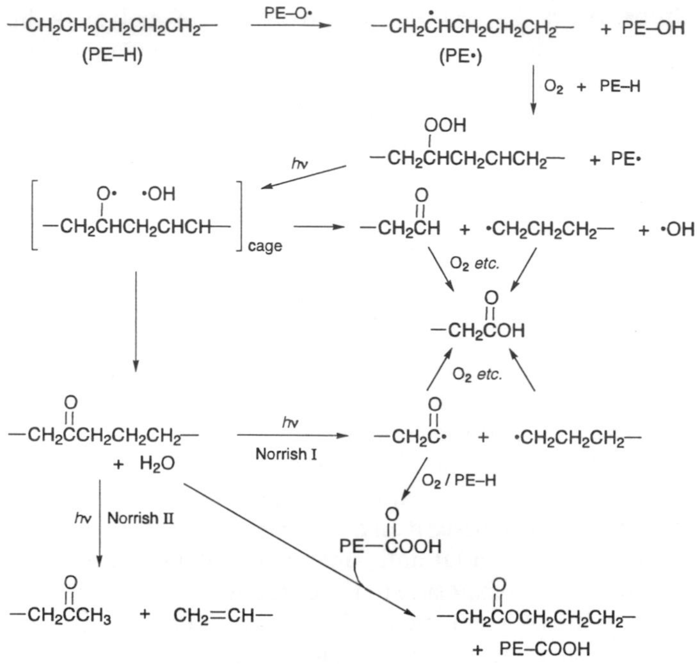

Chapter 3

$-CH_2CH_2CH_2CH_2CH_2-$ $\xrightarrow{\text{PE-O·}}$ $-CH_2\overset{\bullet}{C}HCH_2CH_2CH_2-$ + PE–OH

(PE–H)   (PE·)

$\downarrow$ O$_2$ + PE–H

$$-CH_2\overset{\overset{OOH}{|}}{C}HCH_2CHCH_2- \text{ + PE·}$$

$\left[-CH_2\overset{\overset{O\cdot}{|}}{C}HCH_2\overset{\overset{\cdot OH}{}}{C}HCH- \right]_{cage}$ $\xleftarrow{h\nu}$ $\longrightarrow$ $-CH_2\overset{\overset{O}{\|}}{C}H$ + $\cdot CH_2CH_2CH_2-$ + $\cdot OH$

$\downarrow$ $\searrow$ O$_2$ etc. $\swarrow$

$$-CH_2\overset{\overset{O}{\|}}{C}OH$$

$\swarrow$ O$_2$ etc. $\nwarrow$

$-CH_2\overset{\overset{O}{\|}}{C}CH_2CH_2CH_2-$ $\xrightarrow[\text{Norrish I}]{h\nu}$ $-CH_2\overset{\overset{O}{\|}}{C}\cdot$ + $\cdot CH_2CH_2CH_2-$

+ H$_2$O

$h\nu$ | Norrish II $\quad\quad$ $\downarrow$ O$_2$ / PE–H

$$PE-\overset{\overset{O}{\|}}{C}OOH$$

$-CH_2\overset{\overset{O}{\|}}{C}CH_3$ + $CH_2=CH-$ $\quad\quad$ $-CH_2\overset{\overset{O}{\|}}{C}OCH_2CH_2CH_2-$

+ PE–COOH

Scheme 3.6 *Photooxidation of polyethylene*

Natural polymers such as cellulose, starch, rubber, wool, silk, *etc.* biodegrade on exposure to the environment, although not all at the same rate. The commercial polyesters and polyamides are essentially inert to microorganisms when first manufactured. The poly(urethanes) are much more biodegradable since they are generally amorphous and as foams they have a high surface area. In addition, many PU foams contain polyether segments that are highly oxidisable (both biotically and abiotically) and they biodegrade readily. Abiotic and biotic oxidation are complementary processes. Unbranched hydrocarbons biodegrade more rapidly than branched-chain hydrocarbons but the latter peroxidise more rapidly than the former in the environment. Polyesters and polyamides hydro-biodegrade whereas the polyolefins do not, but the latter photo-biodegrade much more rapidly than the former. These examples illustrate the versatility of the ecosystem to deal effectively with man's pollution problems.

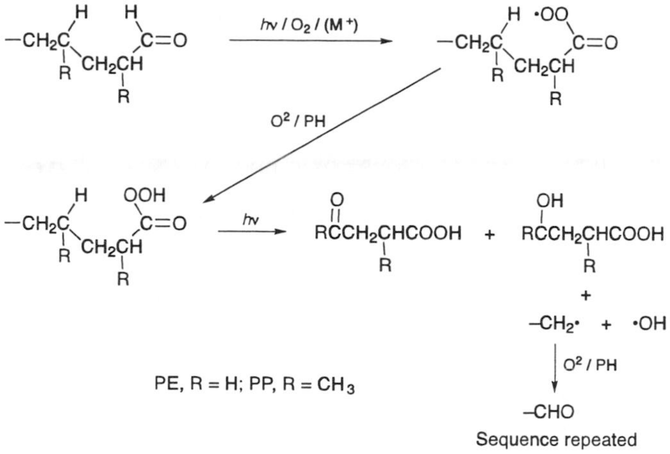

$$
\begin{array}{c}
\text{H} \quad \text{H} \\
-\text{CH}_2\text{C} \underset{R}{\underset{|}{\diagup}} \overset{}{\diagdown}\text{C}=\text{O} \\
\text{CH}_2\text{CH} \\
R
\end{array}
\quad \xrightarrow{h\nu\,/\,O_2\,/\,(M^+)} \quad
\begin{array}{c}
\text{H} \quad \cdot\text{OO} \\
-\text{CH}_2\text{C} \underset{R}{\underset{|}{\diagup}} \overset{}{\diagdown}\text{C}=\text{O} \\
\text{CH}_2\text{CH} \\
R
\end{array}
$$

$O^2\,/\,PH$

$$
\begin{array}{c}
\text{H} \quad \text{OOH} \\
-\text{CH}_2\text{C} \underset{R}{\underset{|}{\diagup}} \overset{}{\diagdown}\text{C}=\text{O} \\
\text{CH}_2\text{CH} \\
R
\end{array}
\quad \xrightarrow{h\nu} \quad
\underset{R}{\underset{|}{\text{R}\overset{\text{O}}{\overset{||}{\text{C}}}\text{CH}_2\text{CHCOOH}}}
\quad + \quad
\underset{R}{\underset{|}{\text{R}\overset{\text{OH}}{\overset{|}{\text{C}}}\text{CH}_2\text{CHCOOH}}}
$$

$$+$$

$$-\text{CH}_2\cdot \;\; + \;\; \cdot\text{OH}$$

$$\downarrow O^2\,/\,PH$$

$$-\text{CHO}$$

Sequence repeated

PE, R = H; PP, R = CH$_3$

Scheme 3.7 *Formation of low molar mass carboxylic acids by environmental peroxidation of polyolefins*

HEAT STABILISATION OF POLYMERS

The term *stabiliser* is the comprehensive technological term used to describe the inhibition of polymer degradation. In practice it applies almost exclusively to the inhibition of peroxidation which, as seen above, can be initiated by mechanooxidation, by light or by transition metal ions. Stabilisers are thus *antioxidants* acting by one or more complementary mechanisms.

The key to the oxidative stability of polymers lies in the control of hydroperoxide formation (reactions 3.1–3.5). This can be achieved either by inhibiting the radical chain reaction or by 'deactivating' hydroperoxides. *Peroxide decomposers* (PDs) remove hydroperoxides without forming free radicals, *metal deactivators* (MDs) are chelating agents that reduce the redox activity of transition metal ions and *UV absorbers* (UVAs) absorb damaging UV wavelengths. These are known collectively as *preventive* antioxidants. In practice hydroperoxides can never be completely eliminated and they are almost always used with antioxidants that interfere with the radical chain reaction, generally referred to as the *chain-breaking* (CB) antioxidants. Preventive and chain-breaking antioxidants acting by different mechanisms complement and reinforce one another by the process known as *synergism*. When synergism occurs between two groups in the same molecule, this is known as *autosynergism*.

Peroxide Decomposers (PDs)

PD antioxidants may destroy hydroperoxides either stoichiometrically or catalytically. The commonly used processing stabilisers, the *alkyl* or *aryl phosphite esters* are representative of the first class:

$$(RO)_3P \ + \ ROOH \ \longrightarrow \ (RO)_3P{=}O \ + \ ROH \qquad\qquad (3.6)$$

However, hydroperoxide decomposers may act by much more complicated mechanisms. Many sulfur compounds, like the *thiodipropionate esters* (DRTPs) or the *metal dialkyldithiocarbamates* (MRDCs) are oxidised to sulfur acids (sulfinic, sulfonic and SO_3) which are ionic catalysts for the non-radical decomposition of hydroperoxides. The MRDCs are particularly important since, unlike the phosphites, they also contain complex transition metal ions and when M is a transition metal ion (*e.g.* Ni) they are also UVAs.

DRTP   MRDC

Hydrogen Chloride Scavengers

A variety of metal salts and oxides have traditionally been used to neutralise hydrogen chloride formed in PVC during processing (page 52). The reaction can be represented generally as:

$$2HCl \ + \ MX_2 \ \longrightarrow \ MCl_2 \ + \ 2HX \qquad\qquad (3.7)$$

Many of the metal salts traditionally used as PVC stabilisers contained heavy metals (*e.g.* lead carbonate, cadmium and barium carboxylates) and although these are cheap and very effective, they are no longer favoured because of their potential toxicological effects in the human environment. Less toxic metal carboxylates (*e.g.* Zn and Ca) are not as effective as Ba and Cd and in recent years attention has turned to the alkyl tin derivatives, notably *dibutyltin maleate* (DBTM) and the *dioctyltin thioglycollates* (DOTGs) which, although more expensive, have the advantage that they are much less toxic and give very transparent PVC for bottles *etc.*

DBTM   DOTG

Most PVC stabilising systems are complex mixtures that contain a variety of additives which complement HCl removal. They include antioxidants to deal with the radicals formed in Scheme 3.5, sulfur compounds to remove hydroperoxides and olefin reactive agents to remove conjugated double bonds which are the source of colour and oxidative instability. Several functions may be found in the same molecule. Thus DOTGs are particularly powerful heat stabilisers for PVC and it has been found that the thiol produced by reaction with HCl has two additional antioxidant functions (Scheme 3.8).

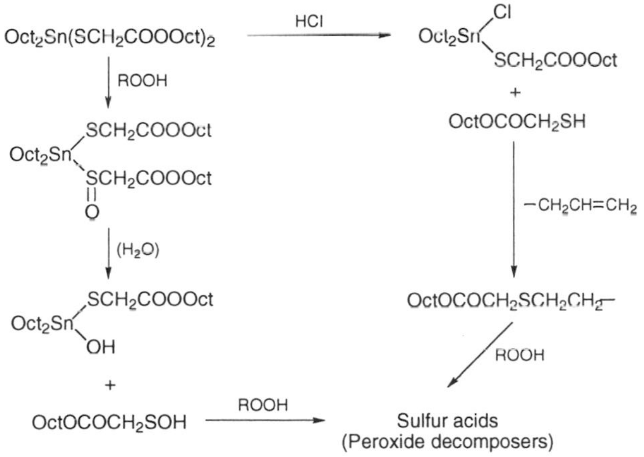

Scheme 3.8 *Autosynergistic mechanisms of DOTG*

The primary mechanism is HCl scavenging, but the tin chloride and the thiol formed from it are hydroperoxide decomposers similar to DLTP described above. However, the liberated thiol can also add to monoenic unsaturation, thus removing reactive allylic groups from the polymer.

Chain-breaking (CB) Antioxidants

Chain-breaking antioxidants (AH) either *donate* a hydrogen atom to peroxyl radicals (CB–D), reaction (3.8), or in their oxidised form (A·) may *accept* a hydrogen atom from an alkyl or peroxyl radical, reaction (3.9):

$$-CH_2\overset{\overset{\displaystyle R}{|}}{C}HOO\bullet \ + \ AH \ \longrightarrow \ -CH_2\overset{\overset{\displaystyle R}{|}}{C}HOOH \ + \ A\bullet \qquad (3.8)$$

$$-CH_2\overset{\overset{\displaystyle R}{|}}{C}H(OO)\bullet \ + \ A\bullet \ \longrightarrow \ -CH=CHR \ + \ AH \ + \ (O_2) \qquad (3.9)$$

Under certain conditions, notably during polymer processing where the oxygen concentration is low, and in light, where the rate of formation of macroalkyl radicals is relatively high, reactions (3.8) and (3.9) may operate together to produce a *catalytic antioxidant* effect.

The most widely used CB antioxidants are the polymer soluble 2,6-*di-tert-butylphenols*, such as BHT and 1076, in plastics and the arylamine antioxidants, *e.g.* PBN and IPPD, in rubbers:

R = CH₃, BHT
R = CH₂CH₂COOC₁₈H₃₇, 1076

PBN   Ethoxyquin

Phenols are less effective antioxidants than the arylamines but they are non-discolouring. They are converted into a variety of oxidation products (*e.g.* quinones and stable phenoxyl radicals) by further reaction with oxygen or peroxyl radicals (page 64). Many of these are themselves antioxidants.

The *N-alkyl-N'-phenyl-p-phenylenediamines* (RPPDs), which are sometimes called 'antidegradants', are unique antioxidants for rubbers because they protect the polymer not only against heat but also against fatigue (mechanooxidation) and 'ozone cracking', all of which are serious problems in tyres. These additives do not behave simply as sacrificial reducing agents for peroxyl radicals, although this is certainly their initial function. It has been shown that some of the oxidation products of the diarylamines, in particular the derived nitroxyl radicals and quinoneimines, are at least as effective as the parent amines. When rubber that contains them is subjected to oscillating stresses (*i.e.* fatigue) they can act catalytically by reactions (3.8) and (3.9), where A· is a *diarylaminoxyl* ($> N–O·$) and AH is the cognate hydroxylamine ($> N–OH$).

LIGHT STABILISATION OF POLYMERS

Stabilisers against the effects of light on polymers are essentially light-stable antioxidants. Many of the peroxidolytic and chain-breaking antioxidants discussed above are relatively ineffective in the presence of light and some, particularly those based on sulfur, are actually photo-prooxidants when used alone. However, they do provide synergistic protection when used in combination with UV absorbers (UVAs).

UV Absorbers and Screens

The earliest light stabiliser to be used in polymers was carbon black, which is still universally used in tyres, at least partly to protect the very peroxidisable rubber molecules from photooxidation. Carbon black is the most effective light absorber known, but it also contains antioxidant groups (phenolic and quinonoid) as part of the polycyclic ring system. In plastics technology, carbon black is normally unacceptable for aesthetic reasons except for specialised applications such as piping. Although not as effective as carbon black, titanium dioxide (a white pigment) is widely used to 'screen' plastics from the effects of UV light out-of-doors, for example in PVC window profiles or cladding. Organic light stabilisers (UVAs) which absorb only in the UV region of the spectrum are colourless and although much less effective than the inorganic pigments as light absorbers, when used in combination with other antioxidants (notably the hydroperoxide decomposers and the hindered phenols) they show powerful synergism. A major application for synergistic combinations of UVAs and antioxidants is in polyethylene films for greenhouses which, as they become more durable and hence cost-effective, are progressively replacing glass. The most important organic UVAs are based on 2-hydroxybenzophenone (*e.g.* HRBP) and 2-hydroxybenzotriazole (HRBT).

HRBP   HRBT

The phenolic UVAs also trap alkoxyl radicals formed from hydroperoxides during photolysis. By removing initiating radical species from the system by reaction (3.8), they are thus also light-stable chain-breaking antioxidants and UVAs. That is they are autosynergistic because they absorb UV light and also act as UV stable CB antioxidants.

Light-stable Peroxide Decomposers

Many PD antioxidants that contain sulfur are ineffective light stabilisers and some even act as photoprooxidants. However, the nickel dithiocarbamates (NiDRC) and the analogous dithiophosphates

(NiDRP) are very effective light stabilisers. They are autosynergists because they are effective UV absorbers and at the same time they catalytically destroy hydroperoxides. NiDNC (in which N denotes isononyl) is used in the light stabilisation of greenhouse films. It will be seen in Chapter 5 that the iron complexes of the dithiocarbamic acids (FeDNC) show short-term light stabilising properties but they *invert* to become prooxidants with time of UV exposure and are used in agricultural mulching films to remove plastics litter from the fields after use.

MDRC   MDRP

Hindered Amine Light Stabilisers

The most effective group of light stabilisers to be developed in recent years are the 'hindered amine light stabilisers' (HALS), based on the 2,2′,6,6′-*tetramethylpiperidine* (TMP) structure. These do not absorb light at all but are oxidised to the corresponding stable *piperidinoxyls* (TMPOs) in the polymer, partially during processing and completely in the presence of light.

(3.10)

TMP   TMPO $\left(\text{>N—O·}\right)$

The TMPs behave similarly to the diaryl nitroxyls in rubber (page 58) and catalytically destroy both macroalkyl and macroalkylperoxyl radicals in the cyclical mechanisms described in reactions (3.8) and (3.9). The aminoxyl (nitroxyl) radicals, $> \text{N–O·}$, trap macroalkyl radicals formed in the polymer and, to complete the cycle, the macroalkyl hydroxylamines, $> \text{N–OP}$, formed are reoxidised to aminoxyl by peroxidic species (such as acylperoxyl and acylhydroperoxides) formed in the polymer (Scheme 3.9).

Spin-trapping agents, *e.g. 2-methyl-2-nitrosopentane* (MNP), which are widely used in chemistry and biology to identify carbon-centred free radicals also react with macroalkyl radicals in polymers during processing to give macroalkyl aminoxyl radicals [reaction (3.11)].

The aminoxyls produced have similar light-stabilising activity to the TMPOs by the catalytic mechanisms shown in reactions (3.8) and (3.9) ($\text{A·} = > \text{NO·}$).

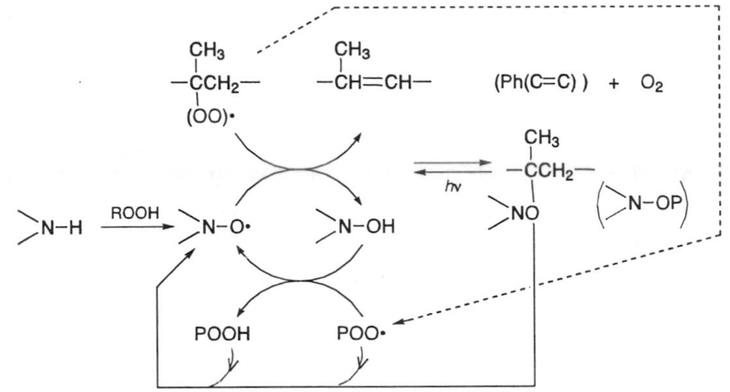

Scheme 3.9 *Catalytic photoantioxidant mechanism of TMPO $\left(\!\!>\!\!N\!-\!O\!\cdot\right)$ in polypropylene*

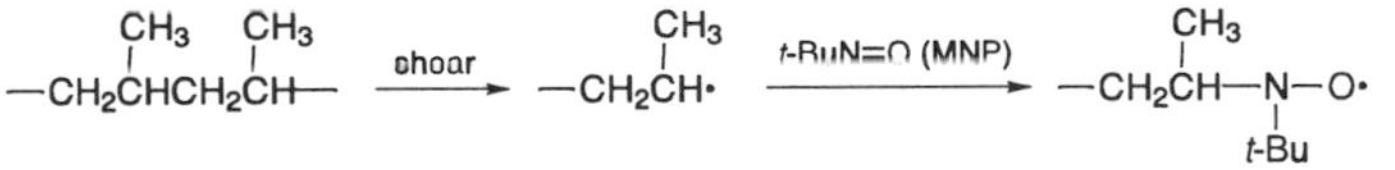

$$-CH_2CHCH_2CH- \xrightarrow{\ shear\ } -CH_2CH\cdot \xrightarrow{\ t\text{-}BuN=O\ (MNP)\ } -CH_2CH-N-O\cdot \qquad (3.11)$$

For further information on antioxidants and stabilisers, the reader is referred to Refs 1–4 and 11–14 at the end of this chapter.

FIRE RETARDANTS

Fire retardant additives are added to polymers during conversion or sometimes during manufacture by chemical reaction with the polymer substrate. Although they retard oxidation under burning conditions, they are not normally antioxidants at ambient temperatures, the mechanisms of action of hydrogen halides in flames are similar to those of antioxidants in the solid phase (see below). Inhibition of gas-phase oxidation is only one of the functions of flame retardants but, since more people are killed in fires by toxic fumes than by flames, it is of paramount importance to inhibit the initiation step in combustion.

Ignition Inhibitors

An important way of inhibiting the ignition of polymeric materials is to increase the formation of carbonaceous 'chars' at the expense of combustible fuels. *Ammonium phosphate* has been used for many years as a flame retardant for cotton and is known to work by catalysing the formation of carbon and water. It is also effective in poly(urethane) foams which form a major component of much domestic upholstery.

Ammonium polyphosphate appears to cross-link the polymer at high temperatures, thus increasing the probability of carbon–carbon bond formation in the condensed phase. There are, however, a number of problems with this method. For example, the temperature at which the polymer begins to degrade is decreased and, because substantial amounts of inorganic salts are required to be effective, they modify the mechanical properties of the PU foam. A particularly disconcerting observation was that under certain conditions toxic phosphorus compounds may be formed with disastrous consequences for human life.

Vapour Phase Retarders

Bromine compounds and to a lesser extent chlorine compounds are known to be inhibitors of the branching radical chain reactions that occur in flames. The most important of these involve oxygen and hydroxyl radicals:

$$H\cdot \ + \ O_2 \longrightarrow \ \cdot O\cdot \ + \ \cdot OH \qquad (3.12)$$

$$\cdot OH \ + \ CO \longrightarrow \ H\cdot \ + \ CO_2 \qquad (3.13)$$

Hydrogen chloride and hydrogen bromide both inhibit these reactions by being converted into the relatively unreactive halogen atoms ($X\cdot$) and may be looked upon as high temperature chain-breaking antioxidants:

$$H\cdot \ + \ HX \longrightarrow \ H_2 \ + \ X\cdot \qquad (3.14)$$

$$\cdot OH \ + \ HX \longrightarrow \ H_2O \ + \ X\cdot \qquad (3.15)$$

Hydrogen halides have to be released over a wide temperature range for maximum effectiveness and this is achieved using the synergist, *antimony trioxide* (Sb_4O_6) which, although ineffective by itself, increases the effectiveness of bromine-containing flame retardants almost three times. It is believed that more thermally stable antimony bromides are formed which partially replace hydrogen bromide in the gas phase:

$$SbBr_3 \ + \ H\cdot \longrightarrow \ SbBr_2 \ + \ HBr \qquad (3.16)$$

Antimony compounds also increase char formation in the condensed phase.

Inert Gas Generators

Water vapour is an excellent flame extinguisher because it vaporises with the absorption of heat and then excludes oxygen from the site of the fire. A similar principle has been developed by incorporating hydrated metal oxides that liberate water under combustion conditions. This has the considerable advantage that the metal oxides themselves, *e.g.* $Al_2O_3 \cdot 3H_2O$ and $Mg(OH)_2$, are inert and the only volatile product is water. However, considerable concentrations (50–60%, w/w) of the water generators are required to give effective blanketing and this reduces the mechanical properties of the composite product, notably strength and impact resistance. This can be improved by increasing the dispersion of the additive in the polymer and its adhesion to the polymer matrix.

ENVIRONMENTAL IMPACT OF POLYMER ADDITIVES

It was realised many years ago that some arylamine antioxidants for rubber were toxic. The best known example is α-naphthylamine, a carcinogen, which was frequently found as a minor impurity (up to 50 ppm) in commercial aryl naphthylamines; notably, the widely used tyre antioxidant *phenyl-β-naphthylamine*, PBN (page 58). A major problem, which initially masked the carcinogenicity of the naphthylamines, was the very long induction period before overt signs of cancer (in this case bladder cancer) began to appear and this required monitoring of workers in the rubber and chemical manufacturing industries over many years. Many process workers in the 1950s and 1960s are still regularly monitored today. Animal tests of course allow toxicity testing to be speeded up but do not have the same degree of certainty as epidemiological studies in humans. It is therefore necessary to be as certain as possible that new polymer additives introduced to the market are 'safe'. Somewhat surprisingly, some arylamine antioxidants, for example ethoxyquin (page 58), are added to animal foodstuffs to preserve them and extensive studies in animals have shown that they have therapeutic activity against some diseases of peroxidation. It is therefore not yet possible to generalise on the basis of chemical structure alone as to which compounds may be toxic. Highly purified PBN has a relatively low toxicity. Other rubber antioxidants, notably the IPPD, have been shown to have skin sensitising (dermatitic) activity in rubber workers but no serious long-term toxicity.

In general the hindered and partially hindered phenolic antioxidants are not toxic to humans and some are permitted as preservatives in fatty foodstuffs (E320). BHT is also used as a processing stabiliser in packaging plastics although, at present, the higher molecular weight and F&DA approved antioxidant 1076 is used more frequently for the same purpose in combination with phosphite synergists.

Antioxidants have to be soluble in plastics to exert their effects. This means that they are also soluble in fats and oils with which the plastics may come in contact, for example in containers or in wrapping films for foods. Consequently, the additives can partition between the polymer and the contacting oils and at least a proportion will 'migrate' into foodstuffs. It was seen above that antioxidants to fulfil their protective function are oxidised to secondary products which, of course, may be subsequently removed. The process of leaching of additives and their transformation products from polymers into the human environment is exacerbated by swelling of the polymer by fats and oils. In general aqueous foodstuffs present less of a problem although dispersed fats, as in milk, may cause some leaching. So far relatively little work has been done to investigate the potential toxicity of additive transformation products although some of them, like the intermediate aldehydes and quinones formed from phenols (Scheme 3.10), are known to be potentially toxic to animals.

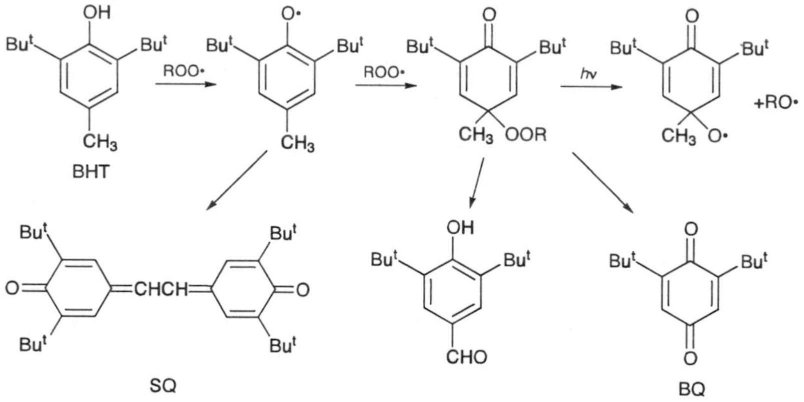

Scheme 3.10 *Oxidative transformation products of BHT*

The fate of the transformation products of PVC additives is a particular cause for concern. HCl scavengers in PVC have to be used at much higher concentrations than conventional antioxidants so that considerable quantities of the corresponding metal chlorides are formed in bottles or films. Furthermore, during the processing of PVC, a substantial proportion of the tin stabiliser is converted into dialkyltin chlorides and maleic anhydride [reaction (3.17)], both potentially toxic products. It is ironic that a great deal of money has been spent by the polymer additive manufacturers to have the parent additives approved by legislative authorities for use in food contact applications, much less attention has been paid, at least in the open literature, to the possible biological effects of the alkyltin chlorides, which are likely to be much more toxic.

$$Bu_2Sn \overset{O}{\underset{O}{\overset{CO}{\underset{CO}{<}}}} \overset{CH}{\underset{CH}{\parallel}} \xrightarrow{2HCl} Bu_2Cl_2 + \overset{HC=CH}{\underset{OC}{\overset{}{\diagup}}} \overset{}{\underset{O}{\diagdown}} CO + H_2O \qquad (3.17)$$

PVC plasticisers have recently come under suspicion as a possible reason for the observed reduction in human sperm count. The causal link between plasticisers and reduced human fertility has not been established to the satisfaction of most epidemiologists and it is essentially based on the theory that the similar shape of plasticiser molecules to the oestrogens allow them to mimic the sex hormones *in vivo*.

POLYMER-BOUND ANTIOXIDANTS AND LIGHT STABILISERS

It was observed in the last section that most antioxidants and stabilisers are relatively low molar mass organic compounds which can be easily extracted by solvents or volatilised from the surface of the polymer at elevated temperature. This is a disadvantage in two ways. Firstly, the additives may migrate into the human environment, causing toxicity in foodstuffs or, as in the case of prostheses, directly in the body. Secondly, because they are lost from the polymer, their effectiveness is reduced.

The latter is a particular problem in hoses, seals and gaskets under the aggressive conditions experienced in engines where they are used in contact with leaching oils and solvents, very often at high temperatures. It has been found necessary to develop special vulcanisation formulations from which the antioxidants are formed as oil-insoluble compounds. A typical example is *basic zinc mercaptobenzothiazolate* (ZnMBT·H_2O) which, like the zinc dithiocarbamates referred to above, is oxidised to sulfur acids which are powerful hydroperoxide decomposers. Unlike the ZnDRCs this compound is completely insoluble and cannot be leached from the rubber. It can thus continue to exert an antioxidant effect for long periods even under severe extraction conditions.

A more versatile method of achieving antioxidant *substantivity* in rubbers is to chemically attach the antioxidant to the polymer. A commercial nitrile–butadiene rubber contains a copolymerised antioxidant, *N-methacrylamidodiphenylamine* (MADPA) which is much more resistant to oil extraction. A similar result has been achieved by forming an adduct of the analogous *mercaptoacryloylamidodiphenylamine*, MADA [reaction (3.18)] by processing of MADA with nitrile–butadiene rubber.

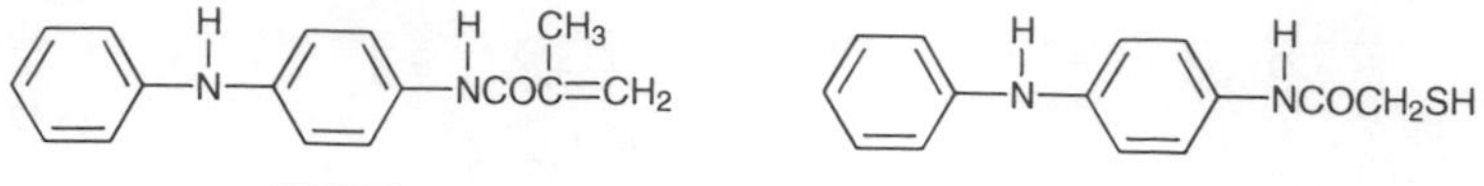

MADPA

MADA

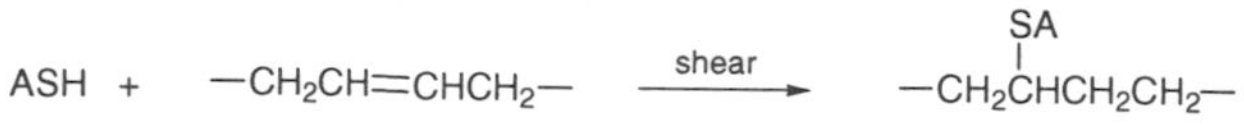

$$ASH \ + \quad -CH_2CH=CHCH_2- \quad \xrightarrow{\text{shear}} \quad -CH_2\overset{\underset{|}{SA}}{C}HCH_2CH_2- \qquad (3.18)$$

Typically, A is a phenolic or arylamine antioxidant (*e.g.* MADA)

For polyolefins used under aggressive conditions (*e.g.* polypropylene fibres subject to dry-cleaning or detergent action) solvent-resistant TMP systems have been developed by covalent chemical attachment of UV stabilisers to the polymer. Two synthetic routes have been developed to do this. The first involves the modification of the polymer (*e.g.* polypropylene) by copolymerisation with maleic anhydride to give an anhydride group in the polymer chain. This can be subsequently reacted with an alcohol or amine group in the light absorber:

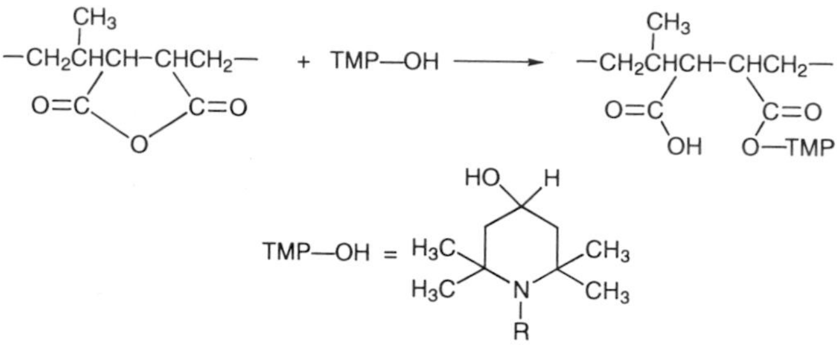

$$(3.19)$$

However, the same effect can be achieved by grafting TMP esters of unsaturated carboxylic acids (*e.g.* BPF) to normal polyolefins by *reactive processing* in the presence of dialkyl peroxides:

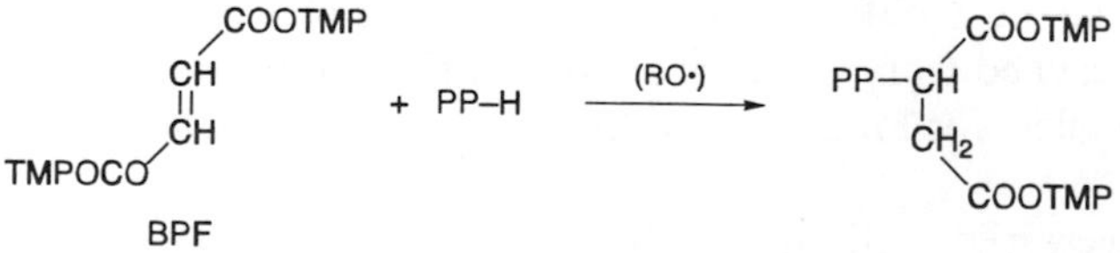

$$(3.20)$$

These modified polymers may be added to the main polymer matrix as a concentrate and the antioxidant cannot be removed from the resulting polymer products by solvent leaching. This makes it possible to wash and dry-clean fibres containing TMP light stabilisers without sacrificing durability.

The rejection of implants by biological organisms is frequently due not to the recognition of the polymer itself as a hostile agent but rather to the slow leaching of toxic chemicals which may include antioxidants and stabilisers. Antioxidant-modified polymers in which the chemical linkage to the polymer is biotically stable provide a potential solution to this problem.

FURTHER READING

1 G. Scott, *Atmospheric Oxidation and Antioxidants*, Elsevier, 1965.
2 N. Grassie and G. Scott, *Polymer Degradation and Stabilisation*, Cambridge University Press, Cambridge, 1985.
3 G. Scott, *Antioxidants in Science, Technology, Medicine and Nutrition*, Albion, Chichester, 1997.
4 G. Scott, Oxidation and stabilisation of polymers during processing, in *Atmospheric Oxidation and Antioxidants*, ed. G. Scott, Elsevier, 1993, Vol. II, Ch. 3.
5 G. Scott, Antioxidants in food packaging: a risk factor?, *Biochem. Soc. Symp.*, 1995, **61**, 235–246.
6 F. Gugumus, The use of accelerated tests in the evaluation of antioxidants and light stabilisers, in *Developments in Polymer Stabilisation*, ed. G. Scott, Elsevier Applied Science, 1987, Ch. 6.
7 G. Scott, Macromolecular and polymer-bound antioxidants, in *Atmospheric Oxidation and Antioxidants*, ed. G. Scott, Elsevier, 1993, Vol. II, Ch. 5.
8 M. M. Herschler, Flame retardant mechanisms, in *Developments in Polymer Stabilisation – 5*, ed. G. Scott, Applied Science Publishers, 1982, Ch. 5.
9 *Polymers and Ecological Problems*, ed. J. Guillet, Plenum Press, 1973.
10 B. B. Cooray and G. Scott, The role of tin stabilisers in the processing and service performance of PVC, in *Developments in Polymer Stabilisation – 2*, Applied Science Publishers, 1980, Ch. 2.
11 G. Scott, Stable radicals as catalytic antioxidants in polymers, in *Developments in Polymer Stabilisation – 7*, Elsevier Applied Science, 1984, Ch. 2.
12 G. Scott, Ageing, weathering and stabilisation of polymers, *Macromol. Chem., Macromol. Symp.*, 1988, **22**, 225–235.
13 G. Scott, Some new concepts in the stabilisation of polymers, *J. Nat. Rubb. Res.*, 1990, **5**, 163–167.
14 G. Scott, Stabilisation of rubber-modified plastics against environmental degradation, in *Developments in Polymer Stabilisation – 1*, Applied Science Publishers, 1979, Ch. 9.

Chapter 4

Management of Polymer Wastes

THE POLYMER WASTE PROBLEM

It was first suggested in the 1960s that the non-biodegradability of synthetic polymers could present a problem in the long term disposal of packaging made from the commodity polymers. Doomsday scenarios began to emerge. Environmental campaigners predicted that by the end of the century the world would be covered by a layer of non-degraded plastic. These predictions failed to take into account nature's diversity, involving both abiotic and biotic chemistry in the bioassimilation of waste and litter (Chapter 3, page 53).

Equally important although not publicised at the time was the accumulation in large dumps of discarded vehicle tyres. These are in fact much more persistent in the environment than plastics and secondary uses have been traditionally found for a proportion of discarded tyres, ranging from impact absorbing buffers on boats and docks, to recreational usage in childrens play areas. Durability has been turned to advantage in these secondary applications of used tyres but in practice they utilise only a small fraction of the polymer wastes and a major proportion of discarded tyres ultimately end up in landfill.

In the 1970s a number of disastrous fires brought the waste tyre problem to the attention of the public. The simplest, although not necessarily the most ecologically acceptable, proposal for the disposal of used tyres was to burn them (with heat recovery). They have a high calorific value and controlled incineration along with removal of sulfur oxides from the effluent permits the recovery of steel tyre-cord as a by-product. Some tyres can be recycled to second usage by re-treading but this requires undamaged carcasses. Recycling by retreading is practiced on a very large scale in some developing countries but it accounts for only a small proportion of used tyres in developed countries since the

Table 4.1 *Generation of post-user plastics waste by end-use in 1995*

End-use	kT	Percentage by weight	Percentage of plastics in total waste
Households	10 139	63.2	7.9
Distribution	2 409	15.0	1.0
Automotive	888	5.5	7.0
Building	841	5.2	0.29
Electrical/Electronic	812	5.1	15.4
Agriculture	293	1.8	0.03
Total	16 056		

Source: Association of Plastics Manufacturers in Europe, APME.

low cost of imported tyres from the developing countries makes retreading economically unfavourable. Rubber 'crumb' from the comminution of waste tyres is being used to some extent in carpet backing and flooring but this again consumes a relatively small proportion of the available waste. Very little attention has so far been paid to the recovery by chemical treatment of the rubber hydrocarbon itself and associated by-products. However, research in Russia has demonstrated that the formation of the polysulfide cross-link is reversible and the rubber produced can be blended with new rubber to give products with very good durability. This example of feedstock recovery points the way forward to the re-use of polymer wastes that are becoming an increasing embarrassment throughout the world.

Over 16 million tons of post-user plastics waste is produced in Western Europe every year and more than half is produced by households (Table 4.1). Although many items of packaging are re-used at least once (notably carrier bags and bottles), this does little to reduce the burden on the municipal waste collection systems. The source of packaging waste is not always clear in published statistics. Stretch-wrap film used for 'packaging' hay is not classified as agricultural waste nor does it appear as industrial waste but it is a severe pollution problem for the farmer and a visual blight in the countryside.

Much domestic packaging is contaminated by heterogeneous materials, particularly residues of foodstuffs, paper labels, metal caps and tags, inks and adhesives, all of which have an adverse effect in mechanical recycling processes. Agricultural packaging is much more seriously contaminated by soil, organic matter and transition metal ions which make it virtually unrecyclable by reprocessing. Contaminants may be partially removed by very thorough washing but this in turn creates its own problems in the disposal of contaminated wash-water, a factor which has to be carefully considered in the overall life-cycle assessment of the product (pages 78–79).

By contrast, relatively uncontaminated single polymer types can be collected from retail outlets by segregation on site (*e.g.* discarded polypropylene crates, battery cases and supermarket shrink-wrap). These can be recycled with the addition of new antioxidants and UV stabilisers into the primary application, generally as a blend with virgin polymer, without significant loss of mechanical properties.

The motor car is a potential source of re-usable waste polymers (*e.g.* from bumpers, battery cases, facias, instrument panels, *etc.*) and these can in principle be returned together with virgin raw material to the original application in a '*closed-loop*'. The primary motivation in recycling battery cases is the recovery of valuable metals but it also offers the additional prospect of returning a proportion of polypropylene to the primary application in a closed-loop. Any polymer component destined for recycling by reprocessing should be initially manufactured with recycling in view. This means that the initial manufacturer must understand how the stabilisation system works to avoid antagonism between the additives. Some car manufacturers are already practicing closed-loop recycling for bumpers. Inevitably some of the original material is lost during service and for this reason alone it is unlikely that the recyclate could ever contain more than 50% of recovered polymer.

An emerging trend in the automotive industry is to minimise the number of engineering plastics to a small core group to facilitate recycling. In addition, if more than one plastic is to be used in any component they must be compatible when re-processed together. Some companies have standardised on a blend of polycarbonate and poly(butylene terephthalate) (PC/PBT) for bumper manufacture. This material is readily recycled but it undergoes hydrolysis during use and the properties are down-graded after recycling. Consequently it cannot be recycled in a closed-loop and it is converted into secondary products which later appear in the waste stream. This is, in principle, a less than ideal solution to the re-use of expensive and high quality engineering polymers.

Large quantities of 'end-of-life' electronic and electrical (E&E) equipment are now beginning to appear in the electronics industries. Microwaves and television sets are the major contributors to this at present but computers may well overtake these due to the very high rate of obsolescence in this industry. The estimated materials value of the materials currently used in the E&E industry, much of it in the form of battery materials, is $\sim$ £50 M per annum and there is a strong incentive to recycle these. The recovery of plastics is generally a by-product of recycling more expensive E&E materials but some success has already been achieved in converting waste plastics into such items as telephone hand sets and circuit boards. However, there is some concern that in this new industry, as is the case in the more mature consumer product industries, that the importance of stabilising polymers for durability may not yet be fully appreciated by recyclers.

The collection and secondary recycling of domestic waste plastics is a much less satisfactory process than in the engineering component industries. The performance of reprocessed plastic bottles generally requires the separation of caps, labels, adhesive tapes, *etc.* If the packaging was designed with recycling in mind, this would not be necessary since the same polymer could be used for caps and labels and adhesives could be avoided by appropriate printing. Unfortunately, the designer of packaging has traditionally been much more interested in the individuality of his product than in its ultimate disposal.

The consumer is often critical about the performance of recycled products. For example, waste collection bags should have the necessary strength and tear resistance to fulfill their essential function but rarely have. It was seen in the last chapter that re-processing of polymers involves mechanooxidation which leads to M_r and mechanical property changes. This could be prevented at small cost by introducing new processing stabilisers during mechanical recycling.

Secondary applications of recycled commodity plastics often involve outdoor uses. There has been a great deal of publicity, particularly in the USA, about the use of mixed plastics waste from packaging to make 'artificial lumber' which is made into park benches, fence posts, cladding for farm buildings. Recycled mixed plastics are also used in such applications as boat docks and fenders which are subject to impact. Toughness retention is therefore essential to the success of recycled materials subjected to mechanical impact during normal service and to have adequate outdoor durability (weathering resistance) effective light stabilisation (Chapter 3) is essential. It is evident already that responsible manufacturers take very seriously the problem of outdoor durability; for example, in the design of rain-water butts, which are now almost entirely manufactured from HDPE packaging waste.

LEGISLATION

In the 1980s the packaging industry was challenged by legislators under pressure from environmental groups to recycle their waste products. The difficulties in doing this successfully with plastics recovered from the waste collection systems were initially greatly underestimated by the industry. Since glass and metals had been recycled for many years, it was claimed by the packaging industry that plastics could be recycled to the same or new products without loss of properties. Legislators took the industry at its word and set high targets for the mechanical recycling of discarded packaging. For example, in Germany the packaging industry was asked to recycle 64% of waste plastics. To date this has not been achieved and it is doubtful if it ever could be within the limits of economic viability. Moreover when plastics waste is recovered from litter the energy utilised in collection and cleansing may be greater than the energy used in the manufacture of the same article from crude oil. There was a much publicised experiment in the USA in which environmentally conscious students were encouraged to collect plastics litter from the sea-shore. It was then transported to a centre many miles away where it was converted into 'plastic lumber' and returned to the sea-shore as seat and benches. History does not record the 'man-hours' used in collecting the waste, nor the energy used in transporting, cleansing and reprocessing it. Suffice it to say that no industrial process has been reported based on the collection of litter.

In spite of the inherent difficulties in the recycling of post-consumer plastics, this is practised to a much higher degree in the developing than in the developed countries. For example, in India there are 18 000 recycling units spread over the whole country, recycling about one million tons of waste plastics each year and employing about 250 000 people. The 'rag-pickers' who collect the materials from rubbish dumps and other highly contaminated sources are mainly children, supported by some women and old men. This industry is extremely profitable to those who run the recycling units and the owners are frequently very rich but there is a high incidence of disease among workers in the industry and the conditions under which it is carried out would not be tolerated in developed countries. Separation is primarily by colour, white being the most valuable, and to a limited extent by polymer type. No attempt is made to separate toxic materials and granulated recyclates of unknown composition are sold on to the primary processors for whatever application they think appropriate. These include milk pouches, footwear, buckets and luggage.

As a result of the higher health and safety standards in developed countries, profitability from recycling is generally much lower and responsibility for packaging waste is now considered to be that of the producer of the packaging rather than the end-user. The *Polluter pays* principle as outlined by the British Government in its consultation paper 'Producer Responsibility for Packaging Waste' states:

Producers of products that end up as waste should take an increased share of responsibility for the recovery of value from those products once they have served their original purpose . . . Producer responsibility gives industry an incentive to make more beneficial use of its own waste and to boost recovery and recycling within its own operations and products.

At the same time, more realistic targets were set for the recovery and recycling of polymer based packaging wastes in the European Union. The 'Waste Framework Directive' put forward by the EU in March 1991 defines recovery as

Recycling/reclamation of organic substances . . . use as fuel to generate energy and spreading on land resulting in benefit to agriculture or ecological improvement including composting and other biological transformation processes

Materials recycling now accounts for only about one-third of all recycled plastics packaging waste. Incineration with energy recovery is currently the major 'recycling' procedure (Table 4.2). Some 75% of the waste comes from domestic sources and the distribution industries (Table 4.1), most of it in the form of packaging that normally ends up in the collected waste systems. However, a proportion of this also ends up in sewage in the form of personal hygiene products, babies 'disposable' nappies, *etc.* which cause considerable problems due to their resistance to biodegradation under these conditions. It will be seen in Chapter 5 that degradable plastics should have an important part to play here in the future.

The EC Directive on *Packaging and Packaging Waste* of 31 December 1994 requires member States to attain the following targets:

(a) By June 30 2001, between 50% and 65% by weight of packaging waste will be recovered.
(b) Within the general target, and with the same time limit, between 25% and 45% by weight of packaging materials in packaging waste will be recycled with a minimum of 15% by weight for each material.

The last phrase is particularly important since, as has already been observed, it is technically much easier to recycle metals, paper and glass than plastics and more energy is vested in the primary production of aluminium, steel, glass and paper than in plastics (Tables 2.1). Moreover, the first three have no value as fuels whereas polymers do (next section). It might be anticipated then that the proportion of plastics ultimately recycled will settle out at the lower end of the range.

In the UK, the packaging industry's response to EU directive has been to set up a number of 'voluntary' bodies to demonstrate the

Table 4.2 *Recovery of post-user plastics wastes. Western Europe 1989–1995*

	Mechanical/ feedstock recycling (%)	Energy recovery (%)	Total plastics waste/ kT	Plastics waste recovered (%)
1989	7.4	14.7	11 433	22.1
1990	7.0	15.5	13 594	22.5
1991	7.4	14.6	14 637	22.0
1992	6.8	15.9	15 230	22.7
1993	5.6	15.0	16 211	20.6
1994	6.3	13.4	17 505	19.7
1995	8.2	16.8	16 056	25.0

Source: APME.

technical feasibility of recycling. Typical of these is RECOUP (RECycling Of Used Plastic containers) which works with the local authorities to collect, segregate and mechanically recycle plastic bottles from domestic sources. Each industrial sector has its own compliance scheme; for example, VALPAK (packaging), DIFPAK (dairy industry) and PAPERPAK (paper industry). The success of each scheme is critically dependent upon the cost of collecting waste, the price of virgin polymers and the quality of the products that can be produced from the waste. Until recently PET was the most 'valuable' waste polymer in the waste stream but with the introduction of new polyester plants, the price of virgin PET dropped rapidly almost to the level of HDPE. This could be equally rapidly reversed in the future.

The efforts of the compliance groups is co-ordinated by an organisation called VALUPLAST, made up of representatives from the raw materials and conversion sectors of the packaging industry. This body aims to prepare business plans to develop increased recovery from the domestic and commercial waste streams. However, it remains to be seen whether appropriate funds will flow back from the conversion end of the industry, who are largely from the traditional polymer processing industries, to allow the local authorities and voluntary bodies to carry out their part hygienically and efficiently.

Plastics litter has been so far largely ignored in legislation, largely because the producing industries do not consider this to be their problem. It is argued that packaging litter is the responsibility of the user and can be eliminated by education and persuasion. However, most of the litter on the sea-shore and in the countryside is not caused by the public but by the shipping, fishing and farming industries. Although contributing less than 2% to the total plastics waste a much higher proportion of agricultural plastics waste (Table 4.1) appears as litter and escapes the recovery net. Plastics are also a much more visible form of environmental pollution than metals, paper or glass, partly because plastics are light and are carried by the wind or float on the surface of water for much greater distances than alternative materials, so that they accumulate in environmentally sensitive areas on the sea-shore and in the countryside. There is still considerable ignorance on the part of municipal authorities about the recycling of plastics. In Florida, a local authority stockpiled 55 000 tons of black plastic mulching film for twelve years because they were told by environmentalists that it was 'black gold' which could be recycled into piping and farm equipment. However, no recycler was prepared to take the material and it had to be landfilled at considerable expense. This material could have been made photo-biodegradable at the manufacturing stage and there are strong environmental arguments for using biodegradable plastics in products that are not readily collected for the alternative recycling procedures (Scheme 4.1, below). The potential for biodegradable plastics in sewage systems, compost and litter will be discussed again in Chapter 5.

DISPOSING OF POST-CONSUMER PLASTICS

Landfill has in the past been the main way of dealing with domestic waste and some industrial packaging. It has a number of important disadvantages.

1. Available 'holes' in the ground have become increasingly scarce and, for political reasons, landfill levies are set to rise sharply in all developed countries during the next decade since municipal waste has to be transported over ever increasing distances with associated wastage of energy and increase in final cost.
2. Domestic and industrial wastes contain a variety of toxic materials (heavy metals and organic chemical wastes) that can leach from unprepared landfills. The modern trend is to line the landfill with an impervious film of plastic (sanitary landfill) to contain water soluble leachates (Chapter 2, pages 30–32).
3. Putrescible residues such as waste biological wastes slowly biodegrade, primarily anaerobically, to give *biogas*. The latter consists mainly of methane, which is 24.5 times more potent than carbon dioxide as a greenhouse gas. It has been estimated by the Organic Recycling and Composting Association (ORCA) that the 12 million tons of methane generated from landfill sites each year has an

Table 4.3 *Calorific values of plastics compared with conventional fuels*

Fuel	Calorific value/MJ kg^{-1}
Methane	53
Gasoline	46
Fuel oil	43
Coal	30
Polyethylene	~ 43
Mixed plastics	30–40
Municipal solid waste	~ 10

effect equivalent to 284 million tons of carbon dioxide. Furthermore, biogas is a frequent cause of explosion and subsidence when houses are built on old landfill sites. However, if it can be generated under controlled conditions and utilised as a fuel it has a higher calorific value than any other common energy source (Table 4.3). Carbon-chain polymers do not biodegrade under anaerobic conditions and plastics are often unfairly criticised for this reason by environmental campaigners. In practice they are much less damaging than paper and other biodegradable materials which should not be included in landfill but should be digested anaerobically to give biogas that can be used as an industrial fuel.

The calorific value of the hydrocarbon plastics is actually higher than that of coal and similar to that of fuel oil (Table 4.3). It is often not fully appreciated by environmental campaigners that the use of waste plastics as fuel places no additional CO_2 burden on the environment since it simply replaces fossil fuels which have already had energy invested in their production. However, it is not correct to call plastics 'renewable resources' unless they are also manufactured from raw materials of biological origin. Landfill should be the last resort for residues that have no fuel or other industrial value; for example, for residues from incinerators after the full fuel potential of the organic materials have been realised. However, these wastes are often highly loaded with toxic heavy metal residues that require careful control and it was seen in Chapter 2 (page 30) that plastics are now playing a very important role as geomembranes in containing dangerous residues so that they do not diffuse into water supplies. A second way of re-utilising the energy of polymeric materials is to pyrolyse them either to useful feedstocks for the polymer industry or as transportable fuel. Scheme 4.1 summarises the complementary waste management options available for polymer-based materials.

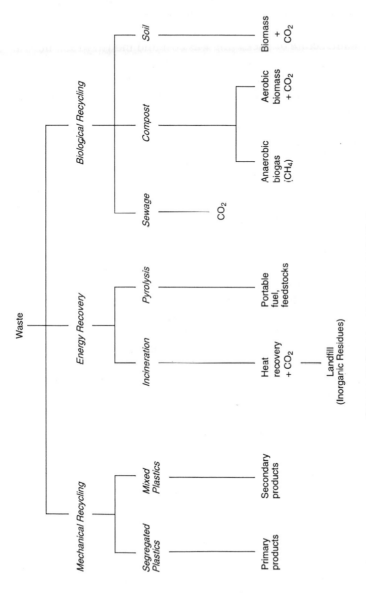

Scheme 4.1 *Polymer waste management options*

LIFE-CYCLE ASSESSMENT

It is now recognised by the manufacturing industry that the design of consumer goods for ultimate disposal introduces a new dimension into its procedures and processes. At one time the only factor to be considered in the manufacture of packaging was cost and industry took no real interest in how its raw materials were produced or how 'society' used and disposed of its products. The regulatory authorities have now made it clear that this is no longer acceptable since manufacturing cost has very little relevance to environmental impact. It is not always clear at first sight which of the alternative disposal systems will minimise energy usage and environmental impact over the life-time and ultimate disposal of the product. The solution to this problem requires a 'cradle-to-grave' assessment of alternative strategies *before* the package is marketed. This has been called 'life-cycle analysis' but this term implies a degree of precision that is not possible in practice. Since the procedure generally involves a comparison of alternative manufacturing and disposal techniques outlined in Scheme 4.1, it is essentially an assessment of the overall energetics and environmental impact of a product. The term *Life-cycle Assessment* (LCA) is now preferred to describe the design of ecologically acceptable packaging. The environmental impacts of packaging materials from the extraction of raw materials (wood, iron, aluminium, oil) to the disposal of waste has been outlined in considerable detail by Bundesamt fuer Umwelt, Wald und Lanshaf (BUWAL) in a report entitled 'An eco-inventory dealing with plastic materials, paper, cardboard, aluminium and glass'. Four environmental parameters, energy consumption, air pollution, water pollution and waste production, are invoked to assess the ecological impact of packaging.

Energy Consumption

Every kind of energy must be considered, of which the most important are electrical, thermal and traction power (MJ) used in the extraction, transformation and transportation of raw materials and in waste treatment to make it safe in the environment.

Air Pollution

The measure of air pollution is the volume (m^3) of pure air that must be added to gas emissions to conform to existing regulations. This includes all gas emissions from raw materials, extraction, production of electrical and thermal energy and in transportation at all stages in the history of the product.

Water Pollution

The measure of water pollution is the volume of pure water (dm^3) that should be used to dilute liquid emissions produced over the whole life-cycle of the product (extraction, production of electricity, *etc.* and in transportation).

Waste Production

In this case the measure is the total volume of wastes (cm^3) produced in each stage in the manufacture, use and disposal of the materials.

The application of the above analysis to the mechanical recycling of plastics packaging is illuminating since it does not always show the environmental benefits predicted by environmental campaigners. There are several contributory reasons for this of which the most important are transport (energy) costs, cleansing (energy utilisation and waste water production), additional polymers or chemicals (energy) to produce a product with performance similar to that from virgin materials. Comparisons with the alternative use of plastics as a fuel generally indicate a positive energy balance for the latter although this must be balanced by the need to eliminate air pollution at some energy cost.

Although LCA is an idealised procedure, it nevertheless provides a basis for comparing the environmental impacts of alternative products and procedures for manufacture to ultimate disposal. In spite of its semi-quantitative nature, LCA provides the manufacturer with information which minimises energy and pollution avoidance costs in his own manufacturing processes. The packaging industry has in the past been less willing to acknowledge the usefulness of LCA in considering final disposal, preferring to believe that domestic waste and particularly litter is the responsibility of the user and the local authorities. It is frequently stated by the packaging industry that litter should be capable of solution by better education of the public! This has been shown to be an unattainable ideal over the past 25 years and, as will be seen in Chapter 5, the biodegradation of agricultural litter can play an integral part in the life-cycle design of products that end up by intention as litter. The following discussion will attempt to apply the principles of LCA to current packaging technology.

MECHANICAL RECYCLING

The basic argument for materials recycling of polymers is that considerable energy has been invested in the raw materials of the polymer

industry and in their conversion into commercial artifacts. An associated argument is that the raw materials resources of the polymer
industry (oil and gas) are limited.

As was noted in Chapter 2 (Table 2.2) energy utilisation during the
manufacture of plastics is substantially lower than for other packaging
materials. Furthermore, fossil carbon resources are considerably greater
than was originally believed. In the 1970s it was estimated that, at the
then rate of usage, oil resources would last for only about 30 years. At
the start of the 21st century it is realised that, owing to advances in the
exploitation of oil resources, reserves are probably an order of magnitude greater than the earlier estimates. In principle this reprieve allows
society to buy time for the development of economically viable alternative technologies for energy generation, leaving fossil resources for
polymer manufacture. Furthermore, some synthetic polymers can even
now be made from waste biological products (*e.g.* ethylene and hence
polyethylene from molasses). It will also be seen in Chapter 5 that
biodegradable plastics [*e.g.* poly(lactic acid) and poly(3-hydroxyalk-
anoates), PHAs] are already being produced from renewable resources.
Their performance is often comparable to commodity plastics but they
have the additional advantage that they are rapidly assimilated into
nature's carbon cycle after discard.

As has already been noted, the arguments for recycling of carbon-
chain polymers are not always clear cut. The motives for materials
recycling may be well meaning but in practice, in developed countries,
the energetics (and hence economics) are often unfavourable. However,
there is a strong argument on ecological grounds for diverting waste
materials from landfill to one or more of the acceptable alternatives
outlined in Scheme 4.1.

The burden for the collection and segregation of used packaging
materials for recycling falls upon the local authority waste collection
centres. These are the 'Cinderellas' of the recycling industry whose work
is rarely recognised in official policy statements. The recycling of plastics
packaging depends on the dedication of poorly paid and often volunteer
individuals who are faced with the unpleasant and frequently un-
hygienic daily task of manually separating bottles and cans contaminated with foodstuff waste (Figure 4.1).

Although an internationally accepted identification code is now used
by the packaging industry to facilitate the separation of one type of
polymer from another this does not make the manual separation task
any more hygienic. The main polymer types can be readily distinguished
by infrared (IR) spectroscopy and landfill levies should be utilised to

Figure 4.1 *A community waste collection centre*
(Photograph courtesy Acre Recycling Centre, Rochdale)

provide low-cost mobile IR facilities to avoid the handling of contaminated packaging by workers.

In the following discussion the materials recycling potential of each of
the main packaging polymers will be considered in the order of decreas-
ing viability.

Poly(ethyleneterephthalate) (PET)

PET is a high-performance polymer originally used in the textile industry. It offers important advantages over PVC in bottles and films
because it is tough, strong and as transparent as glass. It is also very
resistant to peroxidation during processing and in use and for this
reason does not require additives which could migrate into the human

food chain. PET is easily distinguishable from most other packaging polymers because of its transparency. Until recently it was a relatively expensive polymer which was normally in demand in the recycling market. However, its viability has been affected by sharp fluctuations in price of the virgin polymer. Recycled PET packaging, as is the case with other recovered packaging plastics, cannot go back into the original packaging application due to the danger of contamination from toxic chemicals that it may have encountered during first use. Recent attempts by soft drinks companies to overcome this problem by sandwiching recycled PET between outer layer 'barriers' of virgin PET have not been entirely successful since low M_r molecules still diffuse through the 'barrier' into foodstuffs that are in contact with the packaging. A maximum 'shelf-life' of one year has been proposed for such packaging.

More successful applications of recovered PET have been in thermoformable sheets for engineering applications, particularly as alloys with other polymers such as polycarbonate. Applications include sports and garden equipment, electric and electronic goods, car bumpers and even safety helmets. Another important application for recovered PET is as fibre for sleeping bags, carpets and general garment insulation. Cost-performance is the driving force in all these applications and virgin polymer price fluctuation is a major factor affecting the viability of many mechanical recycling processes.

An alternative recycling strategy for PET is to hydrolyse the waste back to the original monomers, giving monomers that can then be sold at stable commodity prices. Re-synthesis of PET gives polymer that can be re-used in its primary packaging applications.

Poly(vinyl chloride) (PVC)

PVC is also widely used in soft drinks and wine bottles, particularly on the European continent. Again it is easily recognisable to those who are responsible for its segregation. Unlike the polyolefins, PVC is heavier than water and this is one method of separating it from other commodity polymers. Unlike PET, PVC has to be formulated with a variety of additives so that it can withstand the processing operation but as discussed in Chapter 3 (page 64) there is some concern about the safety of the processing transformation products formed from PVC stabilisers. These are primarily metal chlorides whose safety in contact with foodstuffs is questionable. Furthermore, successful recycling requires additional stabilisers and it is essential that the provenance of the waste is known since different kinds of PVC stabiliser may be mutually antagonistic, leading to lower stability. *Closed-loop* recycling into the original application may resolve this technical problem but there remains the problem of 'build-up' of stabiliser transformation products which can migrate into the human environment. Recycled PVC can then only be used in down-graded products such as 'disposable' garden trays and plant pots where toxicity is relatively unimportant. PVC is incompatible with the hydrocarbon polymers even in small proportion, causing mechanical weakness in recycled products (see below).

Polystyrene (PS)

PS in pure form is a glassy brittle polymer which is used primarily in display packaging where good transparency is required. It is a relatively minor component of the waste stream in this form. However, its graft copolymers with butadiene (*high impact polystyrene*, HIPS) and with other monomers (ABS, SAN, *etc.*, see Chapter 1, page 12) are widely used in food packaging (cartons *etc.*). They are generally not very durable materials due to their ease of peroxidation and consequently they fragment relatively rapidly.

An important use of PS is in foams and beads which are widely used as impact absorbing materials in packaging. Because of its very low bulk density, expanded PS is a very visible component of plastics litter and is often seen floating on the surface of water. For the same reasons it is particularly expensive to collect and reprocess. Biodegradable foamed plastics based on starch are being developed which promise to be an ecologically acceptable alternative to PS (Chapter 5).

PS possesses a major advantage over PVC and the polyolefins in that it can be pyrolysed (depolymerised) back to styrene monomer in reasonably good yield (42–45%).

Polypropylene (PP)

PP is used in impact modified forms in films, bottles and crates and in unmodified form in ropes and twines. It is frequently present in domestic and industrial waste plastics in association with the polyethylenes. Owing to its pendant methyl groups PP is more readily peroxidised than the other polyolefins and it is not normally separated from them in domestic packaging waste.

Major applications of polypropylene are in milk and beer crates, lettuce bins, pallets, flower buckets, fish trays, ammunition cases, waste bins, *etc.* Many of these are important contributors to coastal pollution. However, because of the weight of polymer used in these items, they are a particularly valuable source of polymer if segregated and delivered to a shore collection point rather than being dumped in the sea. Most of the degradation due to weathering occurs in the surface layers of pigmented polyolefins, leaving the bulk unchanged. The mechanical properties of reprocessed PP crates are consequently not very different from the original and, if the waste is appropriately re-stabilised, it can be blended with virgin polymer to give products with similar durability to the original.

Polypropylene, in addition to its use in plastics is also used in ropes and twines. They frequently end up as litter in the countryside, in the sea and on the sea-shore but they can also be made biodegradable by controlled photoperoxidation (Chapter 5). Considerable amounts of polypropylene are also used in automobile applications and car manufacturers have demonstrated that they can be recycled to the original application if formulated for recycling initially. This means that not only should the component be easily recoverable from the obsolete vehicle but it must be returned to the original manufacturer of the component for appropriate re-stabilisation and blending with virgin polymer. The same stabilisation system, or at least one that is compatible, must be used in the recycling process since the additional additives must not antagonise those originally added (Chapter 3).

Polyethylene (PE)

PE is the major packaging plastic and is the main constituent of waste packaging. All three of the main PE types, LDPE, HDPE and LLDPE (Chapter 1), are used in packaging and for many secondary applications there is no need to distinguish them for the purposes of mechanical recycling. However, for some durable secondary applications, such as water butts, HDPE is segregated and stabilised against environmental degradation.

The effects of minor proportions of PVC and PS in the polyolefins are quite dramatic. As little as 5% of PVC or PS in LDPE reduces the impact strength (toughness) of the latter by about 65%. This results from their presence as separate phases in the polyolefin matrix which leads to rapid crack propagation on impact. The effect of PP is very much less; 10% of PP in LDPE reduced the energy absorbing capacity (toughness) of the matrix by only 1% and 20% of PP reduced it by only 5%. The addition of block copolymers which act as '*compatibilisers*' or more correctly *solid-phase dispersants* (SPDs) for a second incompatible phase reduces the size of the heterogeneous domains and improves impact resistance. However, a considerable concentration ($\sim 20\%$) of SPD is required, which unacceptably increases the cost in most cases.

This problem can be partially overcome by high-shear mixing of a number of polymers by some of the procedures discussed in the next section. High shear mixing of polymers leads to the breaking and reformation of chemical bonds in the polymer backbone and to the *in situ* formation of block copolymers that act as SPDs.

$$-A-A-A-A- \xrightarrow{\text{shear}} -A-A\cdot + \cdot A-A-$$
$$ \rightarrow -A-A-B-B- \quad \text{Block copolymer}$$
$$-B-B-B-B- \xrightarrow{\text{shear}} -B-B\cdot + \cdot B-B-$$

However, as a result of side-reactions of the above macroradicals with oxygen (Chapter 3, page 47 *et seq.*) the durability of polyblends made from recycled plastics is generally inferior to virgin polymers and they are normally used in downgraded, secondary applications.

REPROCESSING OF MIXED PLASTICS WASTES

By far the major proportion of plastics packaging that ends up in a collected waste system is a mixture of the main polymer types discussed in the previous section. Numerous claims appeared in the technical press, particularly in the USA, during the 1980s for the use of these materials. Landfill was the main means of disposing of mixed plastics packaging waste at that time and much of this work was funded by governments in universities, their role being to demonstrate the industrial potential of recycled feedstocks. Since recovered plastics were generally contaminated with materials other than polymers, it was not possible to reprocess them in conventional equipment and entrepreneurs were quick to see the opportunity to invent new procedures for dealing with them. The following processes were developed for the commercial reprocessing of mixed (sometimes called 'co-mingled') waste plastics and some are still in use.

Reclamat Process

This process which was used by Plastics Recycling Ltd. in the UK in the 1980s can accept mixtures of any plastics and a small proportion of non-plastic materials (*e.g.* paper and metals). It does not involve a screw extruder and instead waste granulated polymers are laid on a black plastic film and covered by an identical film to form a sandwich. The mixed plastics are sintered by passing through an oven and finally compression moulded into a board. The product is very tough but has poor dimensional stability under load. Its main application, under the name 'Tuffbord', has been in agricultural structures where its non-absorbent, weather-proof qualities are important. It was later taken over by Superwood Ltd. of Ireland, who re-named it 'Stokbord'. It is particularly useful for protecting underground channels and pipe-work from impact from farming machinery.

Reverzer Process

This system developed by Mitsubishi Petrochemical Company in Japan consists of a modified deep cut extruder designed to give very high shear and hence efficient mixing. In its simplest form, the homogenised material is formed into appropriate shapes by 'flow moulding' in a continuous sequence of metal moulds which are filled in turn. In another version of the Reverzer, the polymer blend is extruded through a large dimension die to give continuous 10×10 cm extrusions.

The basic Reverzer concept has been extended in a more sophisticated injection moulding process to give pallets, stadium seats and cable drums. The outdoor durability of this type of product has not been reported.

Remaker Process

The Remaker, built by Kleindiant in Germany, employs an intrusion moulder. The screw-plasticised mixture is injected at a relatively low pressure through a large diameter gate into aluminium moulds mounted on a conventional press. This machine is very effective for small flexible items but cannot handle mixtures containing PVC because of the relatively long 'dwell time' in the machine.

Flita Process

Designed by Flita Gmbh in Germany, this machine employs a roll mill, open to the atmosphere and eccentrically mounted in a cylindrical mixing chamber. The plasticised mixture is compression moulded into the desired shapes such as perforated paving blocks for car parks. Again durability of these products has not been reported but they would not be expected to be very weather resistant.

Klobbie Process

Invented by E. J. C. Klobbie of Lankhorst Toawfabriken of the Netherlands this machine can handle a mixture of waste plastics with a minimum content of 50% polyolefins and up to 20% PVC and 20% paper (including plastic-coated paper). It uses an adiabatic extruder operating at very high screw speeds and hence shear rate and achieves very good degree of blending of the components. The plastics mixture is extruded at low pressure directly into an open-ended mould. Blowing (expansion) agents are also added to ensure filling of the mould. Typical end products manufactured by Superwood Ltd. include fence posts, agricultural flooring, and Danelaw Ltd. produce underground chambers for the water industries; a very appropriate application since it eliminates weathering, the main weakness of products made from polyblends.

A common feature of all the products made by the above processes is that they depend for their success in having substantial dimensions. Mixed plastics are of little value in the manufacture of films but are a satisfactory replacement for wood concrete and similar materials of construction where the dimensions are sufficient to accept both loads and impacts. A major application for such materials in recent years has been in underground chambers used by the water distributing companies to house stop-cocks. However, for above-ground applications, the high shear used in the manufacture of benches and other products made from 'artificial lumber' leads to loss of mechanical properties during weathering. There is considerable concern therefore about the out-door durability of such products as pallets, park benches and boat docks which are continually subjected to strain and impact over a considerable period of exposure to sunlight.

ENERGY RECOVERY BY INCINERATION

In some countries, notably Japan, incineration is the major method of disposing of combustible waste products from domestic sources. Japan has over 5000 incinerators and most municipalities have their own installations. Until recently, there appeared to be little public protest about this disposal technique in Japan but recent reports have caused some concern about the environmental safety of incineration in heavily populated areas of Japan. Dioxin levels in the air are three times higher than in the USA and in some European countries. Perhaps more significantly, infant mortality rates are 40–70% higher downwind of Japanese incinerators than elsewhere and there is concern about the increase in dioxins in human milk.

On ecological grounds it is also important that the heat produced by incineration of waste polymers (Table 4.3) should be used for the generation of power. It is increasingly evident that this should be carried out locally to reduce transport costs. Both heat and power should also be used locally, thus reducing heat and electricity losses. This points to small but efficient combined heat and power (CHP) incinerators close to centres of population which both produce electricity and provide neighbourhood heat from the waste steam.

The thermoplastic polymers are difficult to burn because they tend to melt and compact. For this reason fluidised-bed furnaces are necessary both to stop compaction and to remove hydrogen chloride and other corrosive gases formed from domestic wastes. These are generally capital intensive and the temperatures must be high enough to destroy potentially dangerous organic compounds such as 2,3,7,8-tetra-chlorodibenzo-*p*-dioxin, one of the most toxic of air pollutants. This aspect of power generation from organic wastes, not only in electrical power generation but also in the production of cement, concerns environmentalists because of the danger of accidental release of toxic chemicals into the environment. Toxic effluents present a particular problem for CHP plants since these operate at maximum efficiency when the heat is produced in or near conurbations, where the escape of toxic gases is most likely to cause maximum damage. Mistrust of the incineration of unknown materials is deep-rooted in parts of Europe and the USA. It seems unlikely then that the incineration of mixed plastics wastes containing chlorinated polymers or halogenated flame retardants will be acceptable to local authorities unless stringent safeguards are built into the system to guarantee that the effluents present no toxic hazards in the event of plant maloperation. Closed-loop recycling (page 82) offers the best prospect of utilising waste chlorinated polymers.

The recovery of energy from municipal solid waste varies enormously across the European Union. Denmark has already achieved a high level of energy generation from waste, whereas the UK has a relatively poor record. This is partly owing to the current economics of landfill but increases in landfill levies will favour energy recovery of energy from waste polymers in the 21st century.

LIQUID FUEL AND FEEDSTOCK RECOVERY

The raw materials of the polymer industry are obtained by *pyrolysis* (or '*cracking*') of oil. In earlier times, they were similarly obtained by pyrolysis of coal, which leads to the formation of coke, tars and 'town gas'. Waste hydrocarbon polymers are similar in chemical structure to mineral oil and on heating to high temperatures they 'crack' to give a mixture of lower molecular weight hydrocarbons (see, for example, Table 4.4), some of which have utility both as chemical feedstocks (*e.g.* the olefins) and the rest as fuels. Pyrolysis can be carried out successfully

Table 4.4 *Composition of gaseous products from the pyrolysis of mixed plastics, wt%*

Product	680	Temperature ($T/°C$) 735	790
Hydrogen	0.667	0.683	1.868
Carbon monoxide	8.376	14.155	6.316
Carbon dioxide	20.418	20.807	3.38
Methane	16.734	22.661	49.491
Ethene	18.383	20.690	25.994
Ethane	10.118	7.189	7.765
Propene	13.758	7.797	3.311
C_3–C_9 hydrocarbons	11.546	6.504	4.875

in fluidised bed furnaces and pilot plants are already in operation (BP, Atochem, DSM consortium). The same kinds of precaution to control the gaseous effluent has to be exercised as in the case of incineration. Some polymers, notably polystyrene and poly(methyl methacrylate), give relatively good yields of the original monomer provided these can be segregated from the polymer wastes.

From mixed plastics, the formation of ethene, methane and hydrogen increase with temperature but the formation of propene and higher hydrocarbons decreases. Whether the isolation of olefins can compete economically with the oil cracking process remains to be seen, but the total hydrocarbon pyrolysate is similar to naphtha, the feedstock for petrochemicals production. It should be noted that because pyrolysis is primarily a non-oxidative process carried out under carefully controlled conditions, the possibility of dioxin formation is much reduced. In 1995 about 100 kT of mixed plastics wastes were converted into chemical feedstocks in Europe by pyrolysis, and this is expected to increase at the expense of incineration over the next ten years.

Two modifications of the pyrolysis route to new feedstocks are currently being explored. The first, developed by Texaco, involves liquefaction and gasification of mixed plastics at high temperatures in the presence of oxygen. The primary product is a high-boiling oil similar to that obtained from coal (~ 70%), liquid hydrocarbons (15%) and fuel gas (2%). The second involves hydrogenation (Veba Oil) which can again utilise mixed plastics wastes.

The hetero-chain polymers, notably the polyesters, polyamides and polyurethanes, can also act as a source of new monomers or oligomers by hydrolysis and alcoholysis. This recycling process (sometimes called 'chemolysis') involves the input of more energy than reprocessing of the recovered plastics but it can, if required, give feedstocks of the same quality as virgin monomers.

MANAGEMENT OF URBAN WASTE

In the last analysis the successful development of ecologically acceptable waste disposal procedures depends on the optimisation of the alternatives to landfill outlined above. The selection of the best combination of the four major alternative options, namely mechanical recycling, composting, energy production and feedstock recovery should minimise adverse environmental impact (including pollution and energy utilisation) as well as financial cost. In practice local authorities take a compliant but pragmatic attitude toward minimisation of landfill to meet current legislation. Perhaps not surprisingly, they emphasise cost minimisation rather than environmental impact in their forward waste disposal planning.

Energy generation by incineration of waste commands a better return than materials recycling and currently plays a major role in local authority plans. Table 4.5 lists the current revenues for sales of recovered materials and Figure 4.2 shows the estimated cost of treating municipal waste over the next 30 years for Greater Manchester.

After 2015 when new incinerators will be in operation and existing contracts for landfill disposal will have run out, landfill will become the most expensive option. Recycling alone is significantly more expensive ($\sim 10\%$) than optimised recycling coupled with energy recovery and a mix of all options gives the lowest cost. Although absolute values would be different for the disposal of plastics waste alone, the cost would be expected to be in the same order. However, this kind of cost calculation does not take into account public acceptance of the alternative scenarios. Materials recycling and composting are probably the top priorities of environmentalists and incineration even with energy recovery almost certainly comes at the bottom of the list. As the public becomes increasingly aware of the effect of air pollution on human health, this order is not likely to change and it seems probable that emphasis will increase in the future on the recovery of plastics materials by both materials recycling and composting. The latter treatment requires biodegradability and it will be seen in the next chapter that even the polyolefins can be made biodegradable through peroxidation.

Table 4.5 *Revenues from recovered waste materials*

Material	Sales income/£ per tonne
Paper	30
Glass	18
Ferrous metals	10
Non-ferrous metals	600
Rigid plastic[a]	100

Source: Courtesy Coopers and Lybrand, Local Government Services 1997.
[a]Recovered from plastics bottles.

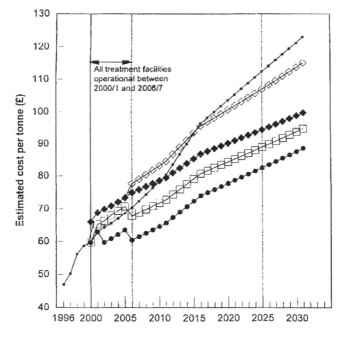

Figure 4.2 *Projected local authority costs of alternative waste management procedures compared with present practice (landfill)* (■) *current position (landfill);* (◆) *maximised mechanical recycling with energy recovery for remaining combustible waste;* (●) *lowest cost waste to energy incineration with mechanical recycling and composting at no extra cost;* (□) *balanced cost mechanical recycling and composting with energy recovery;* (◇) *maximised mechanical recycling and composting without energy recovery* (Reproduced with permission from *Local Government Services; Greater Manchester Waste Disposal Authority Integrated Waste Management Strategy*, Coopers and Lybrand, 1997)

FURTHER READING

1 W. Kaminsky, J. Menzel and H. Sinn, Recycling of plastics, *Conservation Recycling*, 1976, **1**, 91–110.

2 Anon, Low-cost materials from scrap, *Plastics*, May 1983, 56–57.

3 C. Sadrmohaghegh, G. Scott and E. Setudeh, Recycling of mixed plastics, *Polym. Plast. Technol. Eng.*, 1985, **24**, 149–185.

4 R. J. Bunyan, Marketable products from plastics waste, *Plastics Recycling '91*, 1991, Paper 22.1.

5 European Union *Waste Framework Directive*, 75/442/EEC and Amendment 91/155/EEC.

6 European Union *Packaging and Packaging Waste Directive*, 94/62/EC. *Producer Responsibility for Packaging Waste*, UK Department of the Environment, May 1995.

7 Plastics Recycling Foundation USA, *Plastics Recycling: A strategic vision*, 1989.

8 Council for Solid Waste Solutions, *From Soda Bottle to Swimming Pool*, Washington D.C., 1989.

9 G. Scott, Recycling of plastics: A challenge to the polymer industries, *International Conference on Advances in the Stabilisation and Controlled Degradation of Polymers*, Lucerne, May 1990, 215.

10 *Recycling of PVC & Mixed Plastic Waste*, ed. F. P. LaMantia, Chem. Tech. Publishing, 1996.

11 R. W. Renfree, T. J. Nosker, S. Rankin, H. Frankel, T. M. Kasternakis and E. M. Phillips, Physical characteristics and properties of profile extrusions produced from post-consumer co-mingled plastic wastes, *ANTEC'89*, 1809–1812.

12 L. M. Warren and R. Burns, Processors make a go of mixed-waste recycling, *Plastics Technol.*, June 1988, 41–63.

13 E. Koester, Friendly old PET vs cheap new PET, *Mater. World*, 1997, **5**, 525–532.

14 T. Books, Post-consumer bumper recycling, *SAE Technical Paper*, 950554, 1995.

15 W. Hoyle and D. R. Karsa, *Chemical Aspects of Plastics Recycling*, Royal Society of Chemistry, Cambridge, 1996.

16 Management Advisory Council, Waste as Fuel Working Party, *Energy from Waste*, HMSO, London, 1979.

17 House of Lords Select Committee on the European Communities, *Air Pollution from Municipal Waste Incineration Plants*, HMSO, London, January 1989.

18 *This Common Inheritance, Britain's Environmental Strategy*, Part IV, Section 14, HMSO, London, September 1990.

19 *Local Government Services: Greater Manchester Waste Disposal Authority Integrated Waste Management Strategy*, Coopers and Lybrand, 1997.

Chapter 5

Biodegradable Polymers

WHAT ARE BIODEGRADABLE POLYMERS?

The term 'biodegradable' has in recent years become part of the 'green' vocabulary. Man-made polymers are often erroneously contrasted with natural polymers because the latter are perceived to be biodegradable whereas synthetic polymers are not. However, this distinction is artificial. For example, synthetic *cis*-poly(isoprene) and natural rubber have the same chemical structure and they biodegrade rapidly at essentially the same rate in the 'raw' state. When transformed into articles of commerce, *e.g.* the automotive tyre, *cis*-poly(isoprene), whatever its source, becomes highly resistant to biodegradation, not because the basic polymer structure has changed but because the incorporation of the complex antioxidant package essential to performance has made the polymer molecule resistant to oxidation. It was seen in Chapter 3 that low M_r naturally occurring oxidation products are formed from *cis*-poly(isoprene) by peroxidation. It is the initial rate of formation of these biodegradable transformation products that determines the rate of bioassimilation and not the attack of microorganisms on the unmodified polymer chain. Conversely carbon-chain polymers are normally resistant to biodegradation until they are peroxidised.

The above views, which are based on experimental evidence (pages 53–55), are often in conflict with the beliefs of environmental campaigners. The popular view is that all synthetic polymers are 'non-biodegradable' in the environment. The following excerpts from a controversial report sponsored by Greenpeace in 1990 entitled 'Breaking down the degradable plastics scam' epitomise popular misconceptions:

'It is fair to say that every pound of plastic that has been produced, if it has not been burned, is still with us'
'Fragmentation (of photodegraded plastics) occurs without any fundamental

change in the chemical composition of the original plastic'
'Linear chain molecules support fungal growth but only if molecular weights are 500 or less'

These misunderstandings have their origins in early studies of biodegradation in which it was shown that pure hydrocarbon waxes do not biodegrade at molecular weights (M_r) above 500 (*i.e.* $> C_{40}$). However, commercial polyolefins are not 'pure' paraffins. After fabrication they contain a variety of oxidation products which increase their susceptibility to peroxidation with associated loss of mechanical performance. For this reason, commercial rubbers and plastics are always formulated with antioxidants to reduce peroxidation (Chapter 3). This makes them resistant to biodegradation during service and it may well be argued that for short-term applications, such as 'disposable' packaging, plastics are over-stabilised. The persistence of plastics packaging in the countryside in the sea and on the sea-shore is a continuing source of pollution and a danger to animals.

Water-soluble non-biodegradable polymers are also widely used in the detergent, mining and water treatment industries. They present a more difficult problem since to avoid pollution of water courses, they have to biodegrade rapidly, but water solubility is not in itself a guarantee of rapid mineralisation in waste water and sewage systems. For example, poly(acrylic acid) is water soluble but is not biodegradable over a short time-scale. There is increasing awareness that if non-biodegradable water-soluble polymers are discharged into the environment, like 'non-biodegradable detergents' in the 1950s and 1960s, they will sooner or later be a cause of pollution.

'THE GREEN REPORT'

In the 1980s, the response of the enterprising companies within the packaging industry to adverse publicity was to introduce new 'biodegradable' packaging materials, generally based on starch-filled polyethylene, which were claimed to 'disappear naturally' when exposed to the environment. Most of the 'green' claims made for the materials were not supported by experimental evidence and this resulted in a great deal of public concern, particularly in the USA, that packaging manufacturers were making irresponsible and scientifically unjustified claims about the environmental fate of starch-filled polyethylene in common items of packaging such as waste bags and carrier bags. In 1990 the National Association of Attorneys General (USA) published 'The Green Report', based on the findings of a Working Party which drew together these criticisms. The salient conclusions of 'The Green Report' were as follows:

1. To advertise polymers as degradable is deceptive unless the conditions are clearly defined.
2. Degradable plastics must be compatible with existing waste management systems.
3. Meaningful research should be carried out into the effects of degradable plastics in the environment.
4. Testing procedures and protocols for degradability should be established.

This was a constructive and far-sighted contribution to the public debate about the value of degradable polymers to industry and society at large since it led to the establishment of scientifically based testing procedures and protocols for degradable materials in specific environments.

The first 'green' criterion is the governing principle for companies wishing to enter the degradable polymers market. To take an example, it is not legitimate to claim biodegradability in sewage if a polymer does not substantially mineralise during the time it is in the sewage treatment plant. Equally importantly, the same criteria should not be applied to polymers used in agricultural mulching films or in packaging which do not end up in sewage but which require a 'safety' period before they begin to biodegrade in the fields or in compost.

The second 'green' criterion applies particularly to packaging materials whose fate may be landfill or compost but which could also find their way into a mechanical recycling system or an incinerator. It is misleading to claim, as some retailers do, that packaging is 'recyclable' if appropriate facilities are not available to convert the wastes into useful products or fuels by one of the procedures outlined in Chapter 4. In principle all thermoplastic polymers are mechanically recyclable but this is not relevant if appropriate technology is not available or if the cost and energy utilised are greater than the energy saved.

Some polymers used in agriculture (*e.g.* mulching films, baler twines, *etc.*) are designed to degrade at predetermined times as part of their primary function. They will therefore not normally appear as recoverable waste in a recycling system or in compost, and their slow bioassimilation along with nature's biological litter is a valid environmental solution. The rate of mineralisation of agricultural waste in sewage or even in compost is therefore not a relevant measure of biodegradability (page 119 *et seq.*).

The third 'green' criterion indicates that the long-term effects of manmade materials in the environment are as important as their initial impact as litter. The use of degradable materials in consumer products or in agricultural waste must not lead to the generation of toxic or otherwise environmentally unacceptable chemicals in the environment. For example, the use of chlorinated polymers as degradable materials must be looked upon with some suspicion unless the fate of the chlorine can be accounted for as non-toxic products.

The fourth 'green' criterion has been addressed internationally by ISO, in the USA by ASTM and in Europe by CEN and the resulting standards will be discussed below (page 120).

There is still considerable debate even among scientists about the meaning of the term 'biodegradable'. Many assume that it means rapid mineralisation exclusively by bacteria or fungi to give only CO_2 and water. A more general scientific definition of a biodegradable polymer is: 'A polymer in which degradation is mediated at least in part by a biological system'.

The key phrase in this definition is 'at least in part'. A major advantage of synthetic plastics used in packaging is that, as manufactured, they are impermeable and inert toward microorganisms. Thus, if the chemical nutrients which they contain are to be made available to the biological cell, they must first be transformed by chemical reaction to low M_r metabolites. Biodegradation of water-insoluble synthetic polymers is normally induced by the abiotic mechanisms of organic and physical chemistry, namely peroxidation and hydrolysis discussed in Chapter 3. In practice all the common synthetic plastics follow the sequence shown in Scheme 5.1, and, depending on the structure of polymer, the abiotic step may be peroxidation, hydrolysis or a combination of both. However, the rate at which polymers progress from macromolecules to bioassimilable low M_r products varies by orders of magnitude, depending on the nature of the polymer and on the environment.

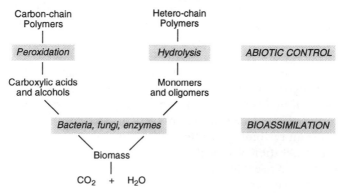

Scheme 5.1 *Biodegradation mechanisms of synthetic polymers*

The degradation environment is critically important to the design of biodegradable polymers. Many waste polymers end up in sewage systems where there is an abundance of microorganisms and rapid hydrolysis or oxidation is the key to their biodegradation. Hydro-biodegradable polymers ideally satisfy this criterion. By contrast, in agricultural applications where there is a cost benefit from the use of biodegradable polymer films, the products have to retain their integrity for weeks or months. Hydro-biodegradation is too unpredictable as an inductive process and antioxidant-controlled peroxidation of hydrocarbon polymers is much more useful where time control is important.

BIODEGRADABLE POLYMERS IN THEORY AND PRACTICE

In principle, all polymers that can be oxidised or hydrolysed should be ultimately biodegradable. However, it may take hundreds of years for bioassimilation to occur. Wood, which is normally considered to be biodegradable, may be highly resistant to biodegradation in some species of tree. A noteworthy example is the sequoia, which contains relatively high concentrations of tannins (hence the red–brown colour of the bark). Fallen sequoia trees are known to have remained intact where they fell for 500 years without microbiological degradation (Figure 5.1). The only deterioration that has occurred over this period is macrobiological erosion at the hands of man. Nature has thus developed an extremely efficient bacteriostat in the tannins. These polyhydroxy phenols not only protect the wood from bacterial and fungal attack but are

Figure 5.1 *Fallen sequoia tree*

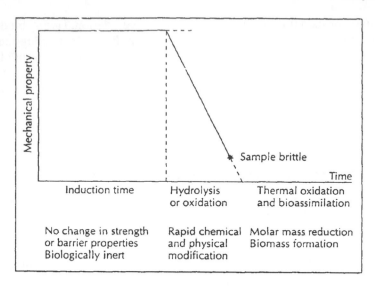

Figure 5.2   *Properties of the ideal degradable plastic*
(Reproduced with permission from G. Scott, *Trend Polym. Sci.*,
1997, **5**, 362)

also very powerful antioxidants with activity similar to the most effective synthetic chain-breaking antioxidants. Because the tannins are chemically linked to the cellulose they are not readily removed physically, and until they are depleted chemically biodeterioration cannot occur.

The ideal behaviour of a degradable polymer used in commercial applications, whether it be natural or synthetic, is illustrated in Figure 5.2. The product must be initially strong and tough, so that it can withstand the stresses imposed on it during service. In the second stage it should physically disintegrate after discard under the influence of the environment and be chemically transformed to carboxylic acids, alcohols, aldehydes and hydroxy acids normally found in nature. In the third stage, the bulk of the polymer should be converted into biomass, CO_2 and water by environmental microflora, thus completing the biological cycle.

Four main types of polymer are currently accepted as being environmentally degradable. They are the photolytic polymers, peroxidisable polymers, photo-biodegradable polymers and hydro-biodegradable polymers. Commercial products may be composite materials in which hydrolysable and peroxidisable polymers are combined (*e.g.* starch–polyethylene composites containing prooxidants). The application, advantages and limitations of each group will be briefly discussed.

Photolytic Polymers

The first degradable carbon-chain polymer was synthesised by Brubaker of the Dupont Company as early as 1950. This was a copolymer of ethylene and carbon monoxide (E–CO) which has since been extensively studied by photochemists, notably by J. E. Guillet and his co-workers at Toronto University. It was seen in Chapter 3 (Scheme 3.6) that macromolecular ketones are formed by peroxidation of polyolefins and by subsequent photolysis they play an important role in the reduction of molecular weight of polyethylene during environmental exposure.

E–CO polymers fragment very rapid in UV light, primarily by the Norrish type II process and the rate increases with the concentration of carbonyl groups. The products of photolysis are vinyl- and ketone-ended polymers which are not rapidly biodegradable and the subsequent biodegradation rate of the relatively high M_r modified polymer is low. The evidence suggests that the partially degraded polymers pass through a M_r minimum ($M_r \sim 15\,000$) with exposure time and then re-polymerise through vinyl and do not biodegrade readily. E–CO polymers are used in packaging where a very rapid rate of fragmentation is required but rapid mineralisation is not important, for example in 'six-pack' collars, which have been reported to entangle animals and birds when carelessly discarded in the countryside or in the sea.

Guillet subsequently synthesised a range of vinyl ketone copolymers with conventional monomers which were marketed as the Ecolyte™ photodegradable plastics. These polymers also photolyse rapidly (Scheme 5.2) but are more biodegradable than the E–CO polymers. Guillet has recently applied the photolysis principle to accelerate the photo-biodegradation of water-soluble polyacrylamide flocculants (Ecofloc™). The initiating step is keto groups randomly distributed along the chain.

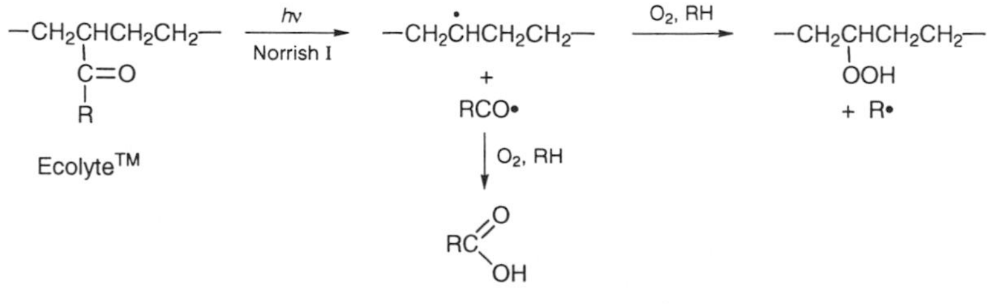

Scheme 5.2 *Photooxidation of ethylene-co-vinyl ketone (Ecolyte™) polymers*

Carbonyl-modified polymers photodegrade too rapidly to be useful in mulching films which normally have to retain their integrity for several months before photodegrading.

Peroxidisable Polymers

Unsaturated carbon-chain polymers (*e.g.* the polydiene rubbers discussed above) are very susceptible to peroxidation and hence biodegradation. Some of these have been studied as photodegradable polymers in their own right. For example 1,2-poly(butadiene) is a plastic with properties similar to the polyolefins. In unstabilised form it photooxidises and thermooxidises rather too rapidly to be very useful commercially.

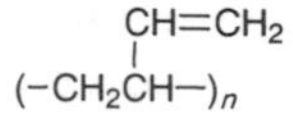

$$CH=CH_2$$
$$(-CH_2CH-)_n$$
1,2-PB

Commercial poly(butadiene), which is mainly the 1,4 isomer, is also used to improve the impact resistance of polystyrene (Chapter 1). Polydienes also increase the rate of physical disintegration of polyblend containing them. The addition of a styrene–butadiene block copolymer (*e.g.* SBS, page 9 *et seq.*) to polyethylene also accelerates the peroxidation of the latter. However, this system also requires a polymer-soluble transition metal ion catalyst (*e.g.* an iron or manganese carboxylate) to increase the rate of photooxidation in the environment by the reactions shown in Scheme 5.3. The products formed by breakdown of alkoxyl radicals (PO·) (Scheme 3.4) are then rapidly biodegradable in compost (page 107 *et seq.*).

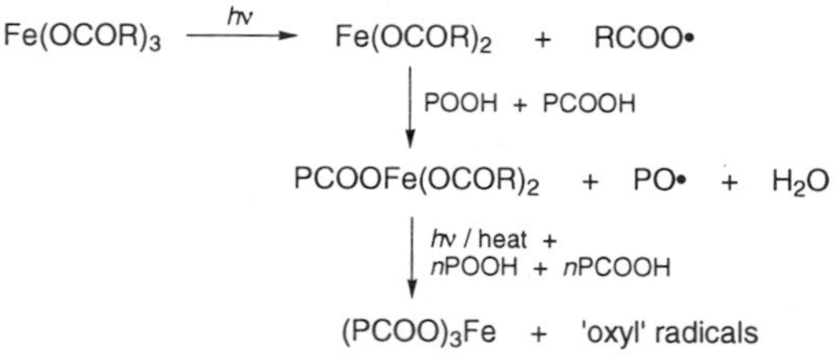

$$Fe(OCOR)_3 \xrightarrow{h\nu} Fe(OCOR)_2 + RCOO\bullet$$

$$\downarrow POOH + PCOOH$$

$$PCOOFe(OCOR)_2 + PO\bullet + H_2O$$

$$\downarrow \begin{array}{l} h\nu \text{ / heat } + \\ nPOOH + nPCOOH \end{array}$$

$$(PCOO)_3Fe + \text{'oxyl' radicals}$$

Scheme 5.3 *Iron catalysed photooxidation of polyethylene*

Transition metal prooxidants cause problems during both the manufacture and use of plastics products. Firstly they catalyse rheological changes in the polymer during processing and reduce shelf-life before use. Secondly it is difficult to control the induction time to photooxidation. It will be seen below that control of peroxidation is essential to the application of degradable plastics in agriculture.

Some organic water-soluble polymers, of which poly(vinyl alcohol), PVA, and poly(ethers) [*e.g.* poly(ethylene glycol), PEG] are the most important, are rapidly peroxidised both abiotically and by microorganisms at ambient temperatures.

$$\underset{\text{PVA}}{(-CH_2\overset{\displaystyle \overset{OH}{|}}{C}H-)_n} \qquad\qquad \underset{\text{PEG}}{(-CH_2CH_2O-)}$$

In the case of PVA, which is one of the few carbon-chain polymers that does not need to be abiotically peroxidised before biological attack, polyketones are the initial products and these are catabolysed to carboxylic acids and bioassimilated by bacteria (*e.g. Pseudomonas*). Kawai has shown (personal communication) that the PEGs behave rather similarly and are bioassimilated by *Sphingomonas*. The active enzymes are believed to be PEG dehydrogenase coupled with cytochrome *c*, the oxidase enzyme of the bacterial respiratory system. Since the polyethers are also very peroxidisable abiotically, abiotic peroxidation may also play a part in the overall process.

Photo-biodegradable Polymers

The biodegradability of carbon-chain polymers is inversely related to the efficiency of the antioxidant systems required to give them durability during manufacture and in service. In the 1960s, studies of the mechanism of peroxidolytic antioxidants (Chapter 3) suggested to the author a potential biostatic control mechanism for the commodity carbon-chain polymers. By manipulation of the antioxidant system it was found that environmental biodegradation could be induced at will by the use of photo-sensitive antioxidants, which protect the polymer during processing and initially during its service life. The transition metal dialkyldithiocarbamates (MDRC), formed in the 'sulfurless' vulcanisation of rubbers, were known to be effective inhibitors of rubber peroxidation during processing and in use. It was seen in Chapter 3 that the essential role of the dithiocarbamates is to destroy hydroperoxides by an ionic mechanism. However, some dithiocarbamates (notably Fe^{III}) are rapidly photolysed by UV light in the outdoor environment. After an induction time that depends on concentration, they sharply 'invert' (Scheme 5.4) with liberation of prooxidant metal carboxylates, resulting in the rapid formation of lower molecular weight carboxylic acids and alcohols by peroxidative scission of the polymer backbone. MDRCs vary profoundly in their stabilising activity in polymers. Thus the nickel and cobalt complexes are effective photoantioxidants which invert to give photoprooxidants.

The key reaction in the time-control step is quantised photolysis or thermal oxidation of the antioxidant system under environmental conditions. Parallel antioxidant and prooxidant reactions occur when the iron dithiocarbamates are exposed to light between 290 and 350 nm.

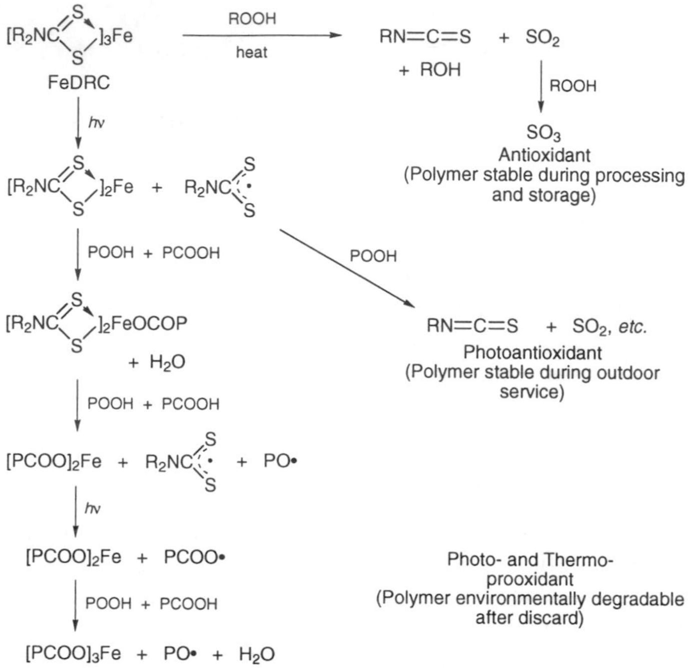

$[R_2NC(S)(S)]_3Fe$ FeDRC $\xrightarrow[\text{heat}]{\text{ROOH}}$ $RN=C=S$ + SO_2 + ROH

$\downarrow$ $h\nu$

$RN=C=S$ + SO_2 + ROH $\xrightarrow{\text{ROOH}}$ SO_3

Antioxidant
(Polymer stable during processing and storage)

$[R_2NC(S)(S)]_2Fe$ + $R_2NC(S)(S)\cdot$

$\downarrow$ POOH + PCOOH

$\searrow$ POOH

$[R_2NC(S)(S)]_2FeOCOP$ + H_2O

$RN=C=S$ + SO_2, etc.

Photoantioxidant
(Polymer stable during outdoor service)

$\downarrow$ POOH + PCOOH

$[PCOO]_2Fe$ + $R_2NC(S)(S)\cdot$ + PO•

$\downarrow$ $h\nu$

$[PCOO]_2Fe$ + PCOO•

Photo- and Thermo-prooxidant
(Polymer environmentally degradable after discard)

$\downarrow$ POOH + PCOOH

$[PCOO]_3Fe$ + PO• + H_2O

Scheme 5.4 *Iron dithiocarbamate (FeDRC) antioxidant-prooxidant inversion in a polyolefin (P represents a polyolefin chain)*

The length of the induction period can be accurately controlled by combining photoprooxidant and photoantioxidant dithiocarbamates.

Photo-biodegradable polyethylene using the above system was developed commercially for use in agriculture by D. Gilead of Plastopil Hazorea in Israel in collaboration with the author and is now widely used as Plastor™ in mulching films in Europe and Plastigone™ in the USA. It is also used in polypropylene baler twines as Cleanfields™ by AMBRACO in the USA, and in controlled release fertilisers as Nutricote™ by Chisso-Asahi Fertilizer Company in Japan. The biodegradation mechanism will be discussed below.

Hydro-biodegradable Polymers

Modified cellulose has been used since 1870 in film-forming materials. Celluloid (nitrocellulose) was followed by cellulose acetate in the 1920s. Water absorption and biodegradability decrease with increasing degree of acetylation which is normally in the region of 40% for packaging films and this material biodegrades slowly. Abiotic processes, particularly hydrolysis and photooxidation also contribute to the ultimate bioassimilation of cellulose acetate and it is much less persistent in the environment than the polyolefins. In the past the main deterrent to the development of biodegradable packaging materials from cellulose has been the traditional method of extracting cellulose from wood. This involved steeping wood pulp in caustic soda, treating it with carbon disulfide, a hazardous chemical, and regenerating it with sulfuric acid. This is a polluting process (Chapter 1) and considerable efforts have been made in recent years to solvent extract the cellulose. The most successful process involves the use of aqueous N-methylmorpholine N-oxide (NMMO). Removal of water leads to the recovery of pure cellulose which can be extruded to give thermoplastic films or fibres (Tensel) which, it is claimed, have initial properties similar to polypropylene but, as might be expected, are much more susceptible to water swelling. Although the solvent can be recovered and the product is highly biodegradable, it is not yet clear whether this 'new' material currently being developed in film form by Viskase Corporation in the USA can compete with the polyolefins in a life-cycle assessment. Academic work is in progress to utilise the lignin component of wood as well as the cellulose to produce composite materials, but the anticipated cost is several times that of the petroleum-based polymers.

Considerable research is also being directed to the synthesis of new polymers which, like cellulose, biodegrade by the hydrolytic route but which can also be fabricated by the standard procedures used in the conversion of thermoplastic polymers (*i.e.* extrusion, injection moulding, *etc.*). Some hetero-chain polymers, notably the polyurethanes, do slowly degrade by microbiological processes and are also readily used as a food source by small mammals and crustacea. The importance of the latter process, which is called *macrobiological degradation* is often overlooked when the ultimate fate of synthetic polymers in the environment is being considered. Small crustacea such as woodlice are found in most environments and they have a voracious appetite for most organic matter. Griffin has reported that the faeces of woodlice contain a proportion of finely divided polyethylene when starch-modified polyethylene is available as a source of food. Macrobiological degradation is disadvantageous in durable consumer goods but it contributes favourably to the physical and chemical modification of man-made polymers before they are ultimately bioassimilated into the natural environment.

A potentially important industrial extension of the production of 'naturally' biodegradable polymers from renewable resources is the utilisation of biological processes to synthesise biodegradable polymers.

These may themselves be utilised after use as a food source by environmental microorganisms, thus incorporating materials technology into
nature's self-contained biological cycle (Scheme 5.5) and potentially
freeing society from its current dependence on fossil fuels as a source of
polymeric materials. This laudable objective is being pursued vigorous-
ly in many academic and industrial laboratories around the world
although, as will be seen below, it brings its own economic and ecological problems.

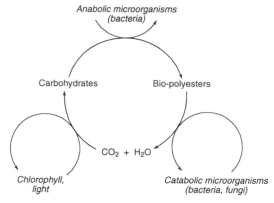

Scheme 5.5 *The extended carbon cycle*

Synthetic hetero-chain polymers, notably the aliphatic polyesters, are
sometimes described as 'truly' biodegradable although the bioassimilation step is normally preceded by abiotic hydrolysis to give monomeric
and oligomeric products that are accessible to microorganisms. There is
no fundamental distinction between biodegradable polymers described
in the previous sections in which biodegradation is initiated by abiotic
peroxidation and the biodegradable polymers which depend on abiotic
hydrolysis as the initiating step. In general the hetero-chain polymers
biodegrade more rapidly than the carbon-chain polymers, but this is not
always the case. Thus, for example, *cis*-polyisoprene, whose biodegradation is initiated by peroxidation, is bioassimilated many times faster
than many polyesters (*e.g.* PET) whose biodegradation is initiated by
hydrolysis.

Typical synthetic biodegradable polyesters that can be made from
renewable resources are poly(glycollic acid) (PGA), commercially available as Dexon™, and poly(lactic acid) (PLA), available as Eco-PLA™,
and their copolymers (*e.g.* Vicryl™):

$$[-O-CH_2-CO-]_n \qquad\qquad [-O-\underset{\underset{CH_3}{|}}{CH}-CO-]_n$$

PGA   PLA

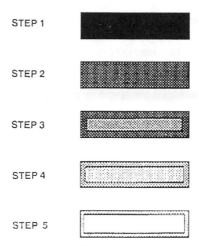

Figure 5.3 *Schematic description of the progress of autocatalytic hydrolysis of*
LA/GA polymers
(Reproduced with permission from S. Li and M. Vert, in
Degradable Polymers: Principles and Applications, ed. G. Scott and
D. Gilead, Chapman & Hall, 1995, p. 54)

Some 80% of the world's production of L-lactic acid is made by the
fermentation of corn sugar (D-glucose) by Purac Biochem BV and the
US consumption for plastics is about 2.5 kT although this is expected to
increase by a factor of three when a new plant operated by Cargill and
Purac comes into operation; Cargill predicts a selling price of
~ \$2.2 kg^{-1}. When made by normal condensation polymerisation,
PLA has a relatively low M_r but this has been improved by Mitsui
Toatsu in Japan to give molecular weights up to 300 000.

The initiating hydrolysis of polyesters occurs more rapidly in the
interior than at the surface owing to autocatalysis by the liberated
carboxylic acid groups which diffuse more rapidly from the surface
layers of the polymer artifact than from the interior. Low M_r carboxylic
acid oligomers diffuse through the undegraded shell and M. Vert and
co-workers have shown that, in the case of GA/LA co-polymers, the
result of this process is a 'shell' of more stable polymer which takes much
longer to biodegrade (Figure 5.3).

The rate of biodegradation is thus very complex and depends not only
on environmental factors but on the physical characteristics of the
polymer artifact, making the overall process difficult to control. Academic and industrial studies are in progress in many laboratories in
Europe, USA and Japan to control-hydrolysis more effectively both for
biomedical devices and for controlled release applications in the body.

Furthermore, if manufacturing costs can be reduced to levels approaching that of the commodity plastics, biodegradable polyesters will also have application in the manufacture of personal hygiene and toilet products and 'disposable' babies nappies which at present accumulate in sewage systems and to some extent on the sea shore as visible and persistent litter. If these consumer products were made biodegradable, this would also open up the possibility of returning them to the biological cycle by controlled composting.

The control of biodegradation rate is of critical importance for many applications of degradable polymers. Amorphous polyesters absorb water and hydrolyse much more rapidly than crystalline materials. Consequently, in partially crystalline polymers, hydrolysis occurs initially in the amorphous phase and continues more slowly in the crystalline phase. This selective degradation leads to an increase in crystallinity by *chemicrystallisation*. A very similar selective abiotic oxidation process occurs in the semi-crystalline polyolefins which fragment rapidly due to failure at the crystallite boundaries.

The bacterially derived polymers of 3-hydroxyalkanoic acids, PHA, of which a copolymer of 3-hydroxybutyric and 3-hydroxyvaleric acids (Biopol™) developed originally by ICI is a commercial example, are relatively crystalline materials whose crystallinity, melting temperature, *etc.* can be varied by increasing the length of the side-chain (R):

$$[-O-\underset{|}{\overset{\overset{\textstyle R}{|}}{CH}}-CH_2-CO-]_n$$

PHA

At present these materials are too expensive to be considered as viable alternatives to the commodity plastics in packaging but they do have potential applications in biomedical products such as orthopaedic implants and even as temporary replacements for parts of the pericardium during open-heart surgery. In this kind of application, performance is much more important than cost. However, Biopol may be able to replace non-biodegradable polymers in paper coating which would then allow paper composite materials to biodegrade much more rapidly in compost and similar environments.

Three main technical weaknesses of the synthetic and biosynthesised polyesters at present limit their usefulness. Firstly that they tend to depolymerise to lower M_r products at normal processing temperatures. This has been partially overcome in the case of the PHAs by reducing the melting temperature (T_m). For example, replacing 60% of 3-hydroxybutyrate (HB) by 3-hydroxyvalerate (HV) in PHB reduces T_m from about 180 °C to just over 100 °C. However, the mechanical performance of PHBV also decreases as the proportion of HV increases. A second disadvantage of the PHAs is that they are biodegraded exclusively by the hydrolytic mechanism. Since this is a random and unpredictable process in the environment, time control is lacking in such applications as agricultural mulching films which have to remain physically unchanged for a pre-determined and reproducible length of time (Figure 5.2). A possible solution to this problem is to combine hydro-biodegradability with time-controlled photo-biodegradability, using light as the quantised initiator of photolysis as described in the previous section.

Finally, and at present crucially, the polyesters are considerably more expensive than the commodity plastics. Work is in progress to genetically modify some plants (*e.g.* rape, *Brassica napus*) to produce seeds containing polyhydroxyalkanoates instead of unsaturated esters. However, it remains to be shown whether polyesters produced in this way can compete on a large scale with the commodity plastics. In typical oilseed plants the yield of oil ranges from 10–50% and the cost of oils from £0.35–0.85 per kg. If 50% yields of PHA could be obtained, prices would be competitive with synthetic plastics. However, it is evident that competition between the use of agricultural land for food and for plastics would become very important if PHAs were to compete with the synthetic plastics in major packaging applications. It has been estimated that at 30% yield of PHA in oilseed crop, 1 Mha (10^{10} m^2), or 7% of the total worldwide area used to grow an oilseed crop, such as rapeseed, would be required to produce 375 000 tons of plastic. This, however, would satisfy only 7% of the US packaging market. It seems doubtful at present whether genetic modification of oilseed plants could provide a viable alternative to conventional polymer synthesis.

Interesting work is currently in progress in the developing countries to use indigenous natural oils (*e.g.* palm oil in Malaysia where there is a potential large excess capacity) or preferably waste biological materials to give PHAs with different kinds of side group, R. Thus for example, the unsaturated fatty esters from palm oil give PHAs which can be readily cross-linked. Another approach under consideration is the genetic modification of tuberous plants to produce PHA. Some 75% of the dry weight of potatoes is starch and if polyesters could be produced in similar quantities, the possibility of producing them at a similar eco-cost to the synthetic polymers would be increased.

Biodegradable Copolymers and Composites

It was noted above that polyurethanes are subject to biodegradation. This principle has been utilised in the development by Showa High

Polymers of Bionolle™ which is reported to be a copolymer containing both ester and urethane groups. The environmental and commercial advantages of this material over conventional degradable polyesters has yet to be demonstrated.

The possibility of combining the physical properties of the polyolefins with hydro-biodegradability of the polyesters has aroused considerable interest among organic chemists. E–CO polymers can be oxidised by hydrogen peroxide by the Baeyer–Villiger reaction to the corresponding esters and this offers a potentially cheap route to polyethylene containing isolated ester groups along the chain. In principle, ester-modified PE should rapidly undergo chain scission at the ester groups to give long-chain carboxylic acids. Information is not currently available as to whether the rate of biodegradation in polymer artifacts is fast enough to be practically useful.

A second group of hydrolysable/biodegradable polymers is based on starch–polymer composites. Starch is cheaper to produce than any of the synthetic polymers and it is highly biodegradable. It has been used for many years as a filler for polyethylene. The addition of corn (maize) starch to polyethylene, pioneered by G. J. L. Griffin of Brunel University, was originally intended to improve the aesthetic properties of polyethylene so that it performed more like paper. However, with the change in emphasis to more biodegradable packaging materials and with the concern that the raw materials for industrial polymers should be made from cheap renewable resources, the latter aspects subsequently dominated commercial development. At the levels used (up to 10%) the polyethylene acts as an encapsulating agent for the starch and only starch accessible to microorganisms in the surface of the polymer is biodegraded. However, biodegradable products were subsequently developed by adding prooxidants (*e.g.* fats, peroxidisable oils or polymers containing unsaturation together with transition metal carboxylates). After embrittlement of the polymer matrix by thermal or photooxidation, release of the biodegradable filler results in rapid bioassimilation. Starch–polyethylene formulations are not very suitable for the manufacture of the very thin films currently used in agricultural mulching films. The reason for this is that most varieties of starch have granule diameters similar to the thickness of the films themselves ($\sim 10\,\mu$m), and this leads to mechanically weak films. The main use of starch-filled polyethylene is to reduce the polyethylene usage (and hence cost) of waste bags which can then be biodegraded in compost. Abiotic peroxidation of the polymer matrix, induced by a transition metal carboxylate, then occurs readily in municipal compost systems and is followed by bioassimilation.

A potentially more versatile approach to starch-based biodegradable polymers has been the development by Novamont of 'thermoplastic starch' in which starch is the major component. Starch can be plasticised by 'extrusion cooking' to give coherent films. However, when water is the plasticiser, the product rapidly hardens to give a brittle material with high tensile strength but little resistance to impact. Hydrophilic plasticisers, which may themselves be polymeric, give much tougher polymer composites. Typical examples of hydrophilic plasticisers used in the commercial manufacture of starch-based polymers, Mater-Bi™, are ethylene–acrylic acid and ethylene–vinyl alcohol copolymers. These composite materials can be readily fabricated and give polymer films that biodegrade in compost and to some extent in activated sludge. The design of such polymer composites involves a compromise between hydrophobic properties which are required for their primary function and hydrophilic properties required for biodegradation. If this balance can be obtained, starch–PE composites would have considerable advantages over traditional polymers in the disposal of personal hygiene products in domestic sewage systems. Like the biodegradable polyesters, the biodegradation of high-starch composites cannot be accurately controlled in the outdoor environment. However, control may be achieved by combining thermoplastic starch technology with degradable polyolefin technology. For example, a reactively processed blend of starch (30%), polyethylene (60%), and antioxidant photoprooxidant and a 'compatibilising' copolymer of ethyl acrylate and maleic anhydride (10%) provides polymer composites whose lifetime can be accurately controlled in sunlight.

AGRICULTURAL APPLICATIONS OF ENVIRONMENTALLY BIODEGRADABLE POLYMERS

One of the most important environmental lessons learned in the latter part of this century is that fresh water is not the prolific and cheap commodity that it has been previously assumed to be. Many of the world's arid regions are well suited to intensive agriculture except for lack of water. Irrigation partly redresses this balance but it is an inefficient process unless positive steps are taken to reduce water loss by evaporation. The use of plastics mulch results in 50% saving of irrigation water and as much as 30% saving in nitrogenous fertilisers even in temperate climates. These saving may be appreciably higher in arid climates and in some desert regions, agriculture can now be carried out successfully on land which was previously barren.

Technical Advantages of Degradable Mulching Films

Plastic mulch performs the same function as straw but does so much more effectively since it creates a closed micro-climate at the plant roots that is not possible with biological materials. This results in earlier cropping (7–12 days) and much increased yields (Chapter 2, Table 2.5).

In the early 1970s the thickness of polyethylene mulching films ranged from 30 to 40 μm. Today, as a result of the pioneering work of Gilead in Israel and China and of Fabbri in Italy using the S–G system, 8–10 μm film is now state-of-the-art. This has resulted in a radical reduction in cost per unit area of mulching film, a factor of major importance to the developing countries. Films of this gauge are too fragile to be collected from the field after cropping and the use of degradable mulching film obviates this necessity. Studies in desert land have shown (Figure 5.4, a–c) that even brackish soils, previously considered to be incapable of supporting agriculture can be cultivated over photo-biodegradable polyethylene. Solar refluxing of pure water from the plastic film results in the soluble salts being carried into the lower layers of the soil where they do not interfere with plant growth. 'Low micron' degradable film technology thus offers the prospect of effectively utilising desert land where agriculture is not normally possible. In more temperate climates where the early part of the year is cold and where water is in increasingly short supply, degradable mulching films are used to increase crop yields and provide an earlier entry to the soft fruit and vegetable markets.

When regular plastics are used in mulching films, the partially de-graded plastic films that remain on the fields have to be removed before the next season. If they are ploughed in they do not fragment and so interfere with subsequent root growth. They have to be removed man-ually at considerable expense to the farmer (Chapter 2, Figure 2.2). Furthermore, many soft fruit crops are now harvested automatically; the fruit being collected and the stems and leaves returned to the ground. The presence of undegraded plastic detritus mixed with the biodegradable residue clogs the machinery and makes automation im-possible. Photodegraded plastic can be processed in the same way as biological waste (see Figure 5.5). Plastor S–G films have the strength and toughness at 10 μm to withstand the stretching and turning under processes at the edges of the film strips during the laying procedure (see Figure 5.6). For the above reasons, very thin photo-biodegradable polyethylene with mechanical properties as good or better than regular PE can now replace the latter in automated harvesting.

The availability of 'low-micron' photo-biodegradable films has also led to revolutionary procedures for growing crops from seed *in situ*. In

(a)

(b)

(c)

Figure 5.4 *Photo-biodegradable polyethylene mulch* (a) *after planting,* (b) *after cropping,* (c) *after ploughing*

Figure 5.5 *Automated harvesting of tomatoes grown on degradable mulch*

Figure 5.6 *Automated laying of degradable mulching film*

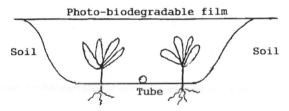

Figure 5.7 *Mid-bed trenching*

the technique known as 'mid-bed trenching', seed is sown directly into a trench covered by degradable plastic, thus creating a 'mini-greenhouse' (Figure 5.7). Photo-biodegradation is timed to match the growth of the plants to the level of the plastic film above them. This procedure not only avoids the cost of transplanting but also eliminates the shock of transplantation and leads to earlier maturity.

Economics of Degradable Mulching Films

Farmers are notoriously conservative and do not take kindly to innovations in agricultural practice, particularly if it is likely to cost more. The key question then in considering the viability of photo-biodegradable plastics is do they improve profitability? Nowhere is cost more important than in China and yet a much higher proportion of the annual output of the polymer industries is used in agriculture in China than anywhere else in the world (Table 5.1) and this usage increased almost 2000 times between 1980 and 1991 (Table 5.2).

However, the critical factor is not how much plastic is used but how efficiently is it used. This depends on the area of ground covered per unit weight of plastic used. Developments in 'low micron' mulching film technology has resulted from improvements in polyethylene technology coupled with the good processing stability of S–G photo-biodegradable polymers. State-of-the-art mulching films are 8–10 μm thick and when this is translated into increase in income to the farmer as a function of cost, all crops show increased profitability compared with conventional

Table 5.1 *Usage of plastics in agriculture (% of total consumption)*

China	20
Israel	12
Spain	8
USA	4

Table 5.2 *Usage of plastic mulch in China*

Year	Hectares ($\times 10^{-3}$)
1980	1.666
1981	21
1983	630
1985	1500
1987	2267
1989	2867
1991	3200

Table 5.3 *Ratio of increased income to cost of mulching film in China*

Crop	Increased income: cost
Melons	13.0
Vegetables	5.0
Peanuts	3.9
Sugar cane	3.6
Cotton	3.0
Maize	2.5

'bare ground' cultivation (Table 5.3). These very thin films are too weak after use to be collected from the soil but their inherent biodegradability removes the need to do so.

Soil Sterilisation

In intensive agriculture, soil pathogens rapidly build up in the soil, leading to serious reductions in crop yields. In the past the favoured method of sterilisation has been treatment of the soil with methyl bromide by injection under plastic films. However, it has been found in Israel that, even with 25 μm polyethylene film, up to 75% of methyl bromide, an ozone-depleting chemical, is lost to the atmosphere within 5 min at 60 °C. International legislation has decreed that the usage of methyl bromide must be reduced to 50% of current levels by 2005 and must be completely eliminated by 2010. Solarisation (ambient sunlight treatment) of the soil covered by photo-biodegradable polyethylene is a practical alternative to chemical sterilisation in areas of high incident sunshine. Gilead has shown that raising the temperature of the soil by means of the Sun's heat alone over a 6 week period kills most of the pathogens while leaving the useful microorganisms. The plastic film is designed to photo-biodegrade at the end of this period.

Controlled Release

S–G photo-biodegradable polyethylene is now being used in a different way to reduce the pollution of water courses by fertilisers. By encapsulating the fertiliser in porous photo-biodegradable capsules fertiliser release times can be achieved from 40 days to one year. Nitrogenous fertiliser based on this principle are manufactured by Chisso-Asahi Fertilizer Company of Japan and scientific studies by Kawai of Okayama University have shown (personal communication) that the empty polymer capsules biodegrade rapidly in soil.

Another potentially very important application of controlled release using photo-biodegradable polyolefins is also being evaluated in the control of disease-bearing insects on the surface of stagnant water. This has particular relevance to the control of the mosquito, the main carrier of the malaria parasite (*Plasmodium*). The infective cycle of this parasite is very extended, taking some months to pass from human blood to the stomach of the mosquito, through the life-cycle of the next generation of mosquitoes and back again to a new human host. Modern biodegradable insecticides do not persist long enough in the aqueous environment to eliminate succeeding generations of the infected insects. It is now possible to extend the release time of the insecticides by encapsulation in photodegradable polyolefins which float on the surface of stagnant water, thus providing a more effective control of this tropical scourge.

Agricultural Packaging

Although originally designed for the control of litter, degradable polyethylene has been slow to take off in packaging applications. Unlike the above uses, there is little incentive to the manufacturer of packaging to use degradable materials. However, there are certain areas of industrial and agricultural packaging where time-controlled degradables are already making a significant contribution to a cleaner and safer environment. An important application in the USA is in polypropylene baler twines which normally remain on the ground after the hay has been consumed by the animals. The use of photodegradable polypropylene, formulated to fragment and biodegrade after one year has resulted in cleaner fields. Similar products have been successfully evaluated in protective bird netting which is again required to last for just one growing season. There is also interest in photo-biodegradable polymers for polypropylene fastenings for vines and even for cartridge cases where a very strict time control is necessary.

An unpleasant and environmentally damaging recent addition to the rural landscape is the wind-borne residues from black stretch-wrap storage bags for hay which are now used in very large tonnages in the UK. Much of the plastic's detritus collects in the fields and festoons hedges and fences after use (Chapter 2, page 22). Thin, black polyethylene can be made photodegradable and the use of photo-biodegradable polymers in environmentally damaging packaging would bring considerable credit to the manufacturers and distributors of agricultural packaging.

BIOASSIMILATION OF PHOTO-BIODEGRADABLE PLASTICS

As seen above, carbon-chain polymers do not normally biodegrade until they are oxidatively transformed into chemical species that can be assimilated by microorganisms. The activity of the scavenging enzymes is not confined to the cell itself and, as was seen above, some dehydrogenases and oxidases are able to catalyse catabolic processes outside the cell. The products of photo-, thermo- and biooxidation are biodegradable low molecular weight carboxylic acids and alcohols, formed primarily in the surface layers of the polymer. It was shown many years ago by Albertsson that the concentration of carbonyl compounds in polyethylene that has been subjected to photooxidation is substantially reduced by exposure to microorganisms. More recently, Albertsson and Karlsson have shown (Ref. 12) that a large number of low molecular weight carboxylic acids formed during photooxidation (see Scheme 3.7) are rapidly removed by microorganisms. Our own studies (Refs. 8–10) have demonstrated that polyethylene film fragments after exposure to soil bacteria for 6 months in the absence of any other source of carbon are deeply eroded with loss of mass (see Figure 5.8). By contrast, sterile (abiotic) culture media have no corresponding erosive effect, indicating that, in the presence of ubiquitous soil microorganisms, the leaching of abiotic oxidation products cannot occur.

Bioerosion occurs without change in molar mass of the bulk polymer, confirming that the microbial attack is initially in the oxidation-modified polymer surface and progression into the polymer depends on the continuation of peroxidation catalysed by transition metal ions. The bacterial colonies also produce oxidase enzymes (*e.g.* cytochrome *P*-450) which produce superoxide and hydrogen peroxide from oxygen of the environment. The latter in turn gives highly reactive hydroxyl

Unoxidised   Photooxidised   Thermooxidised

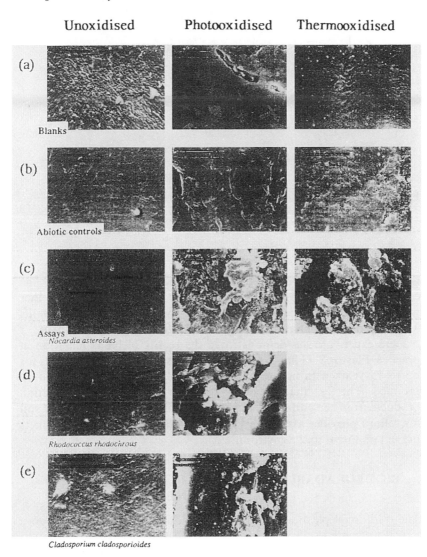

(a)

Blanks

(b)

Abiotic controls

(c)

Assays
Nocardia asteroides

(d)

Rhodococcus rhodochrous

(e)

Cladosporium cladosporioides

Figure 5.8 *Bioerosion of photo-biodegradable polyethylene (Plastor S–G) films after exposure to oxidising environments. SE micrographs (× 5000) obtained after microbial incubation (6 months) in the absence of any other source of carbon*
(Reproduced with permission from R. Arnaud *et al., Polym. Deg. Stabil.,* 1994, **46**, 221)

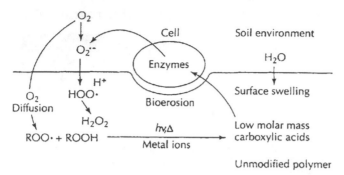

Figure 5.9   *Initiation of aerobic biodegradation in hydrocarbon polymers*
(Reproduced with permission from G. Scott, *Trends Polym. Sci.*,
1997, **5**, 365)

radicals by the Fenton reaction in the presence of iron (Figure 5.9). The
combination of abiotic and biotic initiation of peroxidation, together
with the macrobiological action by insects, worms and animals, results
in the rapid bioassimilation of polyolefins into the soil environment.

Field experience has demonstrated the reality of bioassimilation in
practice. Plastor S–G has been in continuous use on the same fields for
up to 15 years with no accumulation of plastic film or degradation
products. The fact that leaching media do not remove degradation
products from the surface of polyethylene film in the presence of soil
microflora provides assurance that polyolefins peroxidation products
present no threat to the environment.

BIODEGRADABLE PLASTICS IN INTEGRATED WASTE MANAGEMENT

The traditional plastics manufacturing industries have so far given a
muted welcome to the commercial introduction of materials with enhanced degradability. When the concept was first introduced, in the
1970s, it was perhaps understandable that the polymer-using industries should view degradable polymers with some suspicion since polymer technologists had long battled with the durability concerns of user
industries (Chapters 2 and 3). Their objection was brought into
sharper focus in the 1980s when the polymer industries finally accepted
the principle of recycling waste plastics. The British Plastics Federation (BPF) which represents manufacturers of plastics materials in its
'Fact Sheet – Degradability' states 'Degradable plastics cannot be re-

cycled because their lifetime span is unknown and so their growth could complicate an already difficult sorting problem for existing plastics reclaimers . . . degradable plastics could detract from development of recycling initiatives involving heat and power generation.' In practice, degradable plastics used in agriculture are designed to have a well-defined life-span to match their use and once they have completed their intended purpose, they are allowed to biodegrade naturally in the fields and are not collected for any industrial recycling procedure. However, if the consumer was intent on recovering the energy value from degraded polyolefins after use, they could be incinerated like any other hydrocarbon polymer or they could be composted along with biological waste. They could even be reprocessed along with non-degradable mixed plastics litter if there was an economic incentive to do so.

We have already seen (Chapter 4) that the mechanical recycling of mixed plastics waste is generally unsatisfactory and energy wasteful. Exceptions to this are in the manufacture of products involving a considerable bulk of plastics which are not exposed to light (*e.g.* underground chambers). Any degradable material that entered this secondary usage would be trapped inside a large volume of non-degradable plastics which already contain a proportion of biodegradable materials such as paper. Like the latter, they would be essentially non-degradable by either hydrolysis or peroxidation. However, for the reasons given above, it is highly unlikely that any significant amounts of degradable materials will enter any of the recycling systems outlined in Chapter 4 (Scheme 4.1). Products designed to end up in sewage will not and should not be recovered for mechanical recycling although some may be incinerated with municipal refuse. On the other hand, there is no reason to suggest that degradable plastics should not be incinerated since, like other plastics, they will produce a similar amount of energy when incinerated and as was seen in Table 4.3 they will almost certainly have a calorific value higher than that of coal. Biodegradable polyethylene waste bags could end up either in municipal incinerators, in sanitary landfill where they will not biodegrade or in compost where they peroxidise and biodegrade naturally.

DEGRADABLE PLASTICS: POLICY AND STANDARDS

It will be evident from 'The Green Report' (page 94) that the development of reproducible test methods for degradable polymers is an essential prerequisite to the manufacture and use of these materials. Standard test methods can be categorised under three headings which differ according to the environment in which the waste products normally end up.

Sewage

A variety of natural and synthetic polymers appear in the sewage systems. Natural polymers have not in the past caused problems and can be taken as standards to be attained by synthetic polymers. Thus cellulose (paper) is frequently used as a control against which man-made products can be measured. Measurements may be made either under aerobic or anaerobic conditions using a variety of microorganisms found in sewage sludge or in some cases using isolated organisms. For sewage disposal there should be no residual plastics or non-biodegradable organic materials. Polymers that contain elements other than carbon, hydrogen and oxygen may not be fully mineralised and the non-toxicity of any intractable materials should be demonstrated. Typical standard test methods are contained in the following:

(1) International Standards Organisation (ISO)
    ISO/DIS 14851 Plastics – Evaluation of ultimate anaerobic biodegradation of plastic materials in an aqueous medium – Method by determining oxygen demand in a closed resperiometer.
    IOS/DIS 14852 Plastics – Evaluation of ultimate aerobic biodegradability of plastic materials in an aqueous medium – Method of analysis of released carbon dioxide.
(2) American Society for Testing Materials (ASTM)
    Standard test methods corresponding to the above developed by ASTM (Sub-committee D-20.96.01) are
    D5271 Activated-sludge-wastewater-treatment system.
    D5210 Municipal sewage sludge system.

Similar standards are now mandated by the European Commission under the auspices of CEN TC 261 (Packaging): SC4 Packaging and the Environment.

Compost

Composting is a valuable means of recovering valuable agricultural biomass from organic wastes. Many of the commercial degradable polymers can be returned to the biological cycle by this means but the time-scale is very different from that demanded for sewage (Figure 5.10). In compost, unlike sewage, the requirement for complete mineralisation

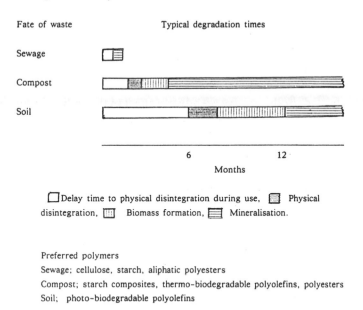

Fate of waste   Typical degradation times

Sewage

Compost

Soil

6   12

Months

☐Delay time to physical disintegration during use, ▨ Physical disintegration, ▦ Biomass formation, ▤ Mineralisation.

Preferred polymers
Sewage; cellulose, starch, aliphatic polyesters
Compost; starch composites, thermo-biodegradable polyolefins, polyesters
Soil; photo-biodegradable polyolefins

Figure 5.10   *Effect of typical biotic environments on polymer bioassimilation*

is not necessary or even desirable since one of the objectives of composting is to produce carbon-based nutrients to act as the seed-bed for new plant growth.

However, there is a tendency both in the USA and Europe to apply the same criteria to both sewage and composting. Draft regulations have proposed that polymers must be 90% converted into carbon dioxide and water during the composting operation (45–180 days). It has been shown by M. Itävaara (personal communication) that many cellulosic materials do not satisfy this criterion, in spite of the fact that cellulose is used as an idealised standard for the biodegradability of packaging materials. The rationale for excluding cellulose from the requirements set for the synthetic polymers is that cellulose has always been part of the human environment whereas synthetic polymers have not and could give rise to toxic products. It was noted above (page 116) that the low molar mass oxidation products of hydrocarbon polymers are rapidly and completely bioassimilated by microorganisms as they are formed by oxidation and cannot be looked upon as a threat to the environment. Similarly, the aliphatic polyesters which hydrolyse to carboxylic acids and alcohols must be similarly innocuous. There is no scientific evidence to suggest that the low M_r products from hydrocarbon polymers and fully aliphatic polyesters are recalcitrant but the situation becomes much less clear when other elements, such as the halogens or heavy metals (*e.g.* from pigments), are introduced into synthetic polymers. The basic requirement of an environmentally acceptable biodegradable polymer in compost is that any by-products that may be formed are either beneficial or harmless in the environment. When elements other than C, H and O are present it should be demonstrated that the end products have no adverse effect in the environment. Typical testing protocols developed for compost are:

(1) ISO
ISO/DIS 14855 Plastics – Evaluation of the ultimate aerobic biodegradability and disintegration of plastics under controlled composting conditions – Method by analysis of released carbon dioxide.
(2) ASTM
D5988-96 Standard test method for determining the aerobic biodegradation of plastic materials or residual plastic materials after composting in contact with soil.
D5247-92 Standard test method for determining the aerobic biodegradability of degradable plastics by specific microorganisms.
D5510-94 Standard practice for heat aging of oxidatively degradable plastics.
D6002-96 Guide to assess the compostability of environmentally degradable plastics.
D5512-94 Standard practice for exposing plastics to a simulated compost environment using an externally heated reactor.
D6003-96 Standard test method for determining weight loss from plastic materials exposed to a simulated municipal solid waste (MSW) aerobic compost environment.

Corresponding European standards are given in WG2 Degradability and compostability. (Similar to D6002-96).

Litter

There are two quite distinct reasons for litter generation. The first and most important is the natural process by which biodegradable materials are recycled in the natural environment. Nature is a prolific litterer and the return of leaves, grass, stalks, twigs and wood to the biological cycle is an essential part of the natural ecosystem. It involves not only microbiological processes but also macrobiological activities by animals below and above the surface of the soil. Human beings play an important role in macrobiological degradation by working the soil surface, which accelerates the process of microbiodegradation.

Man-made litter is more aesthetically objectionable to humans than biological waste. The reasons for this are deeply rooted in the human subconscious because it is evidently not 'natural'. Plastics' detritus is much more noticeable in fields and on the sea-shore and, even when it is ploughed into the soil, it interferes with plant growth.

Biodegradable plastics used in agriculture (page 110) are designed to be integrated into the bio-cycle and, provided they satisfy the tests suggested below, they cause no problems in the environment. Non-biodegradable packaging litter presents a different problem since it is a long-lived and generally very visible by-product of human activity. Most of the plastics packaging litter in the countryside and on the sea-shore is generated not by local populations or by visitors but by commercial enterprise and, in keeping with the 'Polluter pays' principle, the producing industries as well as the using industries should accept a share of the responsibility for its disposal. Most of the industrial plastic detritus comes from the agricultural, shipping and fishing industries. In the UK, about 20 000 tons per annum of polyethylene are used in stretch-wrap packaging for hay. In the early 1990s, 'Farmfilm' was set up in the UK, subsidised by agricultural film producers, to collect and reprocess waste stretch-wrap film from farms. This failed after a few years, mainly because it was not supported by all film manufacturers. Reasons put forward for this were that it was too expensive to transport and it could not be mechanically recycled because of contamination. At present only a small proportion is being recycled.

It is the policy of the British Plastics Federation ('Fact Sheet – Degradability') that 'degradability does not help in the problem of short-term (plastics) litter'. It is concluded in 'Plastics in our lives' (BPF) that 'litter is not caused by packaging but by people; only a constant programme of education to dispose of waste responsibly will reduce the problem.' This ideal, which was proposed first in the 1970s, has not been supported by subsequent experience. In spite of all the endeavours of the 'Tidy Britain Group' commercial packaging litter in the countryside has escalated in the past ten years as the advantages of plastic materials in agriculture and fishing packaging have been exploited but there are no requirements for discarded material to be degradable in the outdoor environment in spite of the concerns of farmers and fishermen themselves about the pollution which they cause.

Not unexpectedly, in the light of the above, there are no published test methods for the biodegradation of plastics litter. The primary requirement is that the detritus breaks down into small particles in a matter of weeks or months, depending on the application (Figure 5.10). For example in the case of discarded six-pack beer collars, this should be a matter of weeks. In mulching films physical breakdown may vary from several weeks to several months and for baler twines, a lifetime of one year is normally required. In the USA, there are standards concerned with the abiotic degradability of six-pack collars but none dealing with their biodegradability. Typical is ASTM's 'D5437-93 Standard Practice for Weathering of plastics under marine floating procedure', which addresses only the first stage of photo-biodegradation. This leaves unresolved the question of whether or how the bioassimilation of oxidation transformation products occurs and whether harmful residues are formed.

Bioassimilation of polymers is primarily controlled by the rate of abiotic peroxidation or hydrolysis. Information is available in the published literature which could form the basis of biodegradation tests but has so far not been incorporated in standard test methods. However, in the long-term it is important that science-based test methods should be developed to permit users of degradable materials in agriculture, shipping and packaging to take advantage of these new materials.

FURTHER READING

1 *Degradable Materials: Perspectives, Issues and Opportunities*, ed. S. A. Barenberg, J. L. Brash, R. Narayan and A. E. Redpath, CRC Press, Boca Raton, FL, 1990.
2 *'The Green Report', Report of a task force set up by the Attorneys General of USA to investigate 'Green Marketing'*, November 1990.
3 *Biodegradable Polymers and Plastics*, ed. M. Vert, J. Feijen, A. Albertsson, G. Scott and E. Chiellini, Royal Society of Chemistry, Cambridge, 1992.
4 *Biodegradable Plastics and Polymers*, ed. Y. Doi and K. Fukuda, Elsevier, 1994.
5 *Degradable Polymers: Principles and Applications*, ed. G. Scott and D. Gilead, Chapman & Hall, 1995.
6 *Chemistry and Technology of Biodegradable Polymers*, ed. G. J. L. Griffin, Blackie Academic & Professional, 1994.
7 G. Scott, Photo-biodegradable plastics: their role in the protection of the environment, *Polym. Degrad. Stabil.*, 1990, **29**, 135–154.
8 S. Williams and O. Peoples, Making plastics green, *Chem. Br.*, 1997, **33**, 29–32.
9 G. Scott, Abiotic control of polymer biodegradation, *Trends Polym. Sci.*, 1997, **5**, 361–368.
10 *Proceedings of the 5th International Scientific Workshop on Degradable Plastics and Polymers*, Stockholm, 9–13 June 1998.

11 G. Swift, in *Kirk-Othmer's Encyclopedia of Chemical Technology*, 4th Edn, Wiley, 1996, vol. 19, pp. 968–1004.
12 *Polymers and Ecological Problems*, ed. J. Guillet, Plenum Press, 1973.
13 A.-C. Albertsson, C. Barnstedt, S. Karlsson and T. Lindberg, Degradation product pattern and morphological changes as a means to differentiate abiotically and biotically aged degradable polyethylene, *Polymer*, 1995, **36**, 3075–3083.
14 R. Arnaud, P. Dabin, J. Lemaire, S. Al-Malaika, S. Chohan, M. Coker, G. Scott, A. Fauve and A. Maaroufi, Photooxidation and biodegradation of commercial photodegradable polyethylenes, *Polym. Degrad. Stabil.*, 1994, **46**, 211.
15 Papers from European Research Society Meeting 1997, Biodegradable Polymers and Macromolecules, ed. A. Steinbüchel, *Polym. Degrad. Stabil.*, 1998, **59**, 3–393.

Subject Index

ABS, *see* Copolymers,
    acrylonitrile–butadiene–styrene
Accelerated UV tests, 42
Additives, 47
Adhesives
    hot melt, 29
    reactive, 29
Agricultural land, use for plastics, 107
Agricultural waste, mineralisation rate,
    95
Agriculture, use of plastics, 24–26
Air pollution, 78
N-Alkyl-*N'*-phenyl-*p*-
    phenylenediamines, 58
Alkyltin chlorides, 64
Americam Society for Testing Materials,
    120, 122
Aminoxyl radicals, 60
Ammonium phosphate, 61
Antifatigue agents, 19
Antimony trioxide, flame retardant, 62
Antioxidant effect, catalytic, 58
Antioxidant–prooxidant inversion, 102
Antioxidants
    arylamine, 58
    chain-breaking, 55, 57, 98
    copolymerised, 65
    diffusion, 32
    durability enhancers, 48
    leaching by body fluids, 35
    loss, 30, 32
    photo-sensitive, 101
    polymer-bound, 65
    preventive, 55
    rubber, dermatitic activity, 63
    substantivity, 32, 65
    volatility, 41

Antiozonants, 19
Arrhenius relationship, 41
Artificial lumber, 87
Autosynergism, 55, 57

Banbury mixer, 49
Battery cases, recycling, 70
BHT, 58
Biodegradability
    definition, 96
    in sewage, 95
Biodegradation
    aerobic, 118
    control of, 106
    mechanisms, 96
Bioerosion, 53, 116
Biogas, 75, 76
Biological cycle, 104
Biomass, from organic wastes, 120
Bionolle™, 108
Biopol, 106
Bladder cancer, 63
Bottles, returnable, 20
Bridge bearings, 29
Bromine compounds, chain reaction
    inhibitors, 62

Carbon black, as light stabiliser, 59
Carbon monoxide, 21
Catalysts, metal ion, 100
Cavity insulation, 29
Cellulose, 2, 3, 4
Cellulose acetate, bioassimilation, 103
CEN TC 261 (Packaging), 120
Chalking, 28
Chemicrystallisation, 106
Chemolysis, 89